Owl 'The Mysterious Bird'

Hiren B. Soni, Ph.D.

ISBN 978-93-5610-610-9
© Hiren B. Soni, Ph.D. 2022
Published in India 2022 by Pencil

A brand of

One Point Six Technologies Pvt. Ltd.
123, Building J2, Shram Seva Premises,
Wadala Truck Terminal, Wadala (E)
Mumbai 400037, Maharashtra, INDIA
E connect@thepencilapp.com
W www.thepencilapp.com

Author biography

Dr. Hiren B. Soni has 24 years of research and 16 years of academic experience in diverse fields such as Systematic Zoology (Invertebrates & Vertebrates), Ornithology & Wildlife Biology, Field Biology (Biodiversity & Conservation), Environmental Impact Assessment (EIA), Wetland Biology (Biodiversity, Ecology, Restoration, Management), Bioremediation (Phytoremediation), and Ecotoxicology. He is credited with more than 150 publications including research reports, research papers, scientific articles, popular publications, book chapters and books at national and international levels. Moreover, Dr. Hiren B. Soni is voluntarily serving as Associate Editor, Certified Reviewer of Publons Academy (UK), Deputy Editor, Editorial Advisory Board Member, Editorial Book Reviewer, Editor-in-Charge, Managing Editor, Manuscript Editor, Panel Reviewer, and Senior Editor in international publishing houses e.g. Springer, Elsevier, Taylor & Francis etc. He is also affiliated with Nature Publishing Group as a Potential Reviewer in Nature Journal (Environment & Ecology), Scientific Data Journal (Ecology, Environment, Biodiversity, Conservation) and Nature Communications (Earth & Environment). Dr. Hiren B. Soni is also acting as Advisory Member (Environmental Consultant) in Hanell International, Zimbabwe (South Africa).

CONTENTS

Preface

Owls are birds from the order Strigiformes, which includes over 200 species of mostly solitary and nocturnal birds of prey typified by an upright stance, a large, broad head, binocular vision, binaural hearing, sharp talons, and feathers adapted for silent flight. Exceptions include the diurnal northern hawk-owl and the gregarious burrowing owl. Owls hunt mostly small mammals, insects, and other birds, although a few species specialize in hunting fish. They are found in all regions of the Earth except polar ice caps and some remote islands. Owls are divided into two families: the true (or typical) owl family, Strigidae, and the barn-owl family, Tytonidae. Throughout human history, owls have variously symbolized dread, knowledge, wisdom, death, and religious beliefs in a spirit world. In most Western cultures, views of owls have changed drastically over time. Owls can serve simultaneously as indicators of scarce native habitats and of local cultural and religious beliefs. Understanding historical and current ways in which owls are viewed, and not imposing Western views on other cultures, is an important and necessary context for crafting owl conservation approaches palatable to local peoples. The Book Owl: The Mysterious Bird is a systematic compilation of authentic and standard literature published by renowned ornithologists, professionals, amateurs and hobbyists of the world in the field of ornithology and wildlife biology. It also covers author's personal observations and published research work in the wild terrains of Gujarat (India) during his research exposure of 24 years as an Ornithologist and Wildlife Biologist. The present book highlights the remarkable key-points and noteworthy information about owls in terms of introductory notes on anatomy, sexual dimorphism, adaptations for hunting, flight and feathers, vision, hearing, talons, beak, camouflage behavior, breeding and reproduction, evolution and systematics, food and hunting, pellets, eyes, plumage, feet, breeding, calls, holistic information about owls (e.g. food and hunting, pellets, eyes, plumage, feet, breeding, calls), symbolisms about owls (African cultures, ancient European and modern Western culture, Asia, Hinduism, native American cultures), motifs and mythology, markers of gods, knowledge, wisdom, and fertility, owls in lore and culture, owls in tribal folklores, owls in mythology and culture, owls in Greek and roman mythology, owls in English folklore, owls in American Indian culture, owl mythology: a global perspective, human impacts on owls, individual impacts, toxicological impacts on owls, agricultural impacts on owls, impacts of habitat alteration on owls, legal protection to owls, owls as environmental indicators, towards a tolerant conservation, rodent control, attacks on humans, and conservation issues. The author has also acknowledged all the researchers, scientists, authors, website developers, NGOS, URLs, and direct or indirect informers in the form of 'Citations' in the section 'Selected Bibliography'. This book will definitely be a ready reference material and handy study guide for students, researchers, scientists, folklore specialists, and bird conservationists around the world. The author renders his immense gratitude and enormous thankfulness to all the informants, who have helped him directly or indirectly during the compilation of this work.

Acknowledgements

To All the Naturalists of The Mother Planet 'Earth' !

Introduction Owls

Owls are birds from the order Strigiformes, which includes over 200 species of mostly solitary and nocturnal birds of prey typified by an upright stance, a large, broad head, binocular vision, binaural hearing, sharp talons, and feathers adapted for silent flight. Exceptions include the diurnal northern hawk-owl and the gregarious burrowing owl. Owls hunt mostly small mammals, insects, and other birds, although a few species specialize in hunting fish. They are found in all regions of the Earth except polar ice caps and some remote islands. Owls are divided into two families: the true (or typical) owl family, Strigidae, and the barn-owl family, Tytonidae.

Anatomy

Owls possess large, forward-facing eyes and ear-holes, a hawk-like beak, a flat face, and usually a conspicuous circle of feathers, a facial disc, around each eye. The feathers making up this disc can be adjusted to sharply focus sounds from varying distances onto the owls' asymmetrically placed ear cavities. Most birds of prey have eyes on the sides of their heads, but the stereoscopic nature of the owl's forward-facing eyes permits the greater sense of depth perception necessary for low-light hunting.

Although owls have binocular vision, their large eyes are fixed in their sockets - as are those of most other birds - so they must turn their entire heads to change views. As owls are farsighted, they are unable to see clearly anything within a few centimeters of their eyes. Caught prey can be felt by owls with the use of filoplumes - hair like feathers on the beak and feet that act as "feelers". Their far vision, particularly in low light, is exceptionally good.

Owls can rotate their heads and necks as much as 2700. Owls have 14 neck vertebrae compared to seven in humans, which makes their necks more flexible. They also have adaptations to their circulatory systems, permitting rotation without cutting off blood to the brain: the foramina in their vertebrae through which the vertebral arteries pass are about 10 times the diameter of the artery, instead of about the same size as the artery as in humans; the vertebral arteries enter the cervical vertebrae higher than in other birds, giving the vessels some slack, and the carotid arteries unite in a very large anastomosis or junction, the largest of any bird's, preventing blood supply from being cut-off, while they rotate their necks. Other anastomoses between the carotid and vertebral arteries support this effect.

The smallest owl - weighing as little as 31 gm and measuring some 13.5 cm - is the Elf owl (Micrathene whitneyi). Around the same diminutive length, although slightly heavier, are the lesser known long-whiskered owlet (Xenoglaux loweryi) and Tamaulipas pygmy owl (Glaucidium sanchezi). The largest owls are two similarly sized eagle owls; the Eurasian eagle-owl (Bubo bubo) and Blakiston's fish owl (Bubo blakistoni). The largest females of these species are 71 cm (28 in) long, have a 190 cm wing span, and weigh 4.2 kg.

Different species of owls produce different sounds; this distribution of calls aids owls in finding mates or announcing their presence to potential competitors, and aids ornithologists and birders in locating these birds and distinguishing species. As noted above, their facial discs help owls to funnel the sound of prey to their ears. In many species, these discs are placed asymmetrically, for better directional location. Owl plumage is generally cryptic, although several species have facial and head markings, including face masks, ear tufts, and brightly colored irises. These markings are generally more common in species inhabiting open habitats, and are used in signaling with other owls in low-light conditions.

Sexual Dimorphism

Sexual dimorphism is a physical difference between males and females of a species. Female owls are typically larger than the males. The degree of size dimorphism varies across multiple populations and species, and is measured through various traits, such as wing-span and body mass. Overall, female owls tend to be slightly larger than males. The exact explanation for this development in owls is unknown. However, several theories explain the development of sexual dimorphism in owls.

One theory suggests that selection has led males to be smaller because it allows them to be efficient foragers. The ability to obtain more food is advantageous during breeding season. In some species, female owls stay at their nest with their eggs while it is the responsibility of the male to bring back food to the nest. However, if food is scarce, the male first feeds himself before feeding the female. Small birds, which are agile, are an important source of food for owls. Male burrowing owls have been observed to have longer wing chords than females, despite being smaller than females. Furthermore, owls have been observed to be roughly the same size as their prey. This has also been observed in other predatory birds, which suggests that owls with smaller bodies and long wing chords have been selected for because of the increased agility and speed that allows them to catch their prey.

Another popular theory suggests that females have not been selected to be smaller like male owls because of their sexual roles. In many species, female owls may not leave the nest. Therefore, females may have a larger mass to allow them to go for a longer period without starving. For example, one hypothesized sexual role is that larger females are more capable of dismembering prey and feeding it to their young, hence female owls are larger than their male counterparts. A different theory suggests that the size difference between male and females is due to sexual selection: since large females can choose their mate and may violently reject a male's sexual advances, smaller male owls that have the ability to escape unreceptive females are more likely to have been selected.

Adaptations for Hunting

All owls are carnivorous birds of prey and live mainly on a diet of insects and small rodents such as mice, rats, and hares. Some owls are also specifically adapted to hunt fish. They are very adept in hunting in their respective environments. Since owls can be found in nearly all parts of the world and across a multitude of ecosystems, their hunting skills and characteristics vary slightly from species to species, though most characteristics are shared among all species.

Flight and Feathers

Most owls share an innate ability to fly almost silently and more slowly in comparison to other birds of prey. Most owls live a mainly nocturnal lifestyle and being able to fly without making any noise gives them a strong advantage over their prey that are listening for the slightest sound in the night. A silent, slow flight is not as necessary for diurnal and crepuscular owls given that prey can usually see an owl approaching. While the morphological and biological mechanisms of this silent flight are more or less unknown, the structure of the feather has been heavily studied and accredited to a large portion of why they have this ability.

Owls' feathers are generally larger than the average birds' feathers, have fewer radiates, longer pennulum, and achieve smooth edges with different rachis structures. Serrated edges along the owl's remiges bring the flapping of the wing down to a nearly silent mechanism. The serrations are more likely reducing aerodynamic disturbances, rather than simply reducing noise. The surface of the flight feathers is covered with a velvety structure that absorbs the sound of the wing moving. These unique structures reduce noise frequencies above 2 kHz, making the sound level emitted drop below the typical hearing spectrum of the owl's usual prey and also within the owl's own best hearing range. This optimizes the owl's ability to fly silently to capture prey without the prey hearing the owl first as it flies in. It also allows the owl to monitor the sound output from its flight pattern.

The feather adaption that allows silent flight means that barn owl feathers are not waterproof. To retain the softness and silent flight, the barn owl cannot use the preen oil or powder dust that other species use for waterproofing. In wet weather, they cannot hunt and this may be disastrous during the breeding season. Barn owls are found frequently drowned in cattle drinking troughs, since they land to drink and bathe, but are unable to climb out. Owls can struggle to keep warm, because of their lack of waterproofing, so large numbers of downy feathers help them to retain body heat.

Vision

Eyesight is a particular characteristic of the owl that aids in nocturnal prey capture. Owls are part of a small group of birds that live nocturnally, but do not use echolocation to guide them in flight in low-light situations. Owls are known for their disproportionally large eyes in comparison to their skulls. An apparent consequence of the evolution of a large eye in a relatively small skull is that the eye of the owl has become tubular in shape. This shape is found in other so-called nocturnal eyes, such as the eyes of strepsirrhine primates and bathypelagic fishes. Since the eyes are fixed into these sclerotic tubes, they are unable to move the eyes in any direction. Instead of moving their eyes, owls swivel their heads to view their surroundings.

Owls' heads are capable of swiveling through an angle of roughly 2700, easily enabling them to see behind them without relocating the torso. This ability keeps bodily movement at a minimum, thus reduces the amount of sound the owl makes as it waits for its prey. Owls are regarded as having the most frontally placed eyes among all avian groups, which gives them some of the largest binocular fields of vision. However, owls are farsighted and cannot focus on objects within a few centimeters of their eyes. These mechanisms are only able to function due to the large-sized retinal image. Thus, the primary nocturnal function in the vision of the owl is due to its large posterior nodal distance; retinal image brightness is only maximized to the owl within secondary neural functions. These attributes of the owl cause its nocturnal eyesight to be far superior to that of its average prey.

Hearing

Owls exhibit specialized hearing functions and ear shapes that aid in hunting. They are noted for asymmetrical ear placements on the skull in some genera. Owls can have either internal or external ears, both of which are asymmetrical. Asymmetry has not been reported to extend to the middle or internal ear of the owl. Asymmetrical ear placement on the skull allows the owl to pinpoint the location of its prey. This is especially true for strictly nocturnal species such as the barn owls Tyto or Tengmalm's owl. With ears set at different places on its skull, an owl is able to determine the direction from which the sound is coming by the minute difference in time that it takes for the sound waves to penetrate the left and right ears. The owl turns its head until the sound reaches both ears at the same time, at which point it is directly facing the source of the sound. This time difference between ears is about 30 microseconds.

Behind the ear-openings are modified, dense feathers, densely packed to form a facial ruff, which creates an anterior-facing, concave wall that cups the sound into the ear structure. This facial ruff is poorly defined in some species, and prominent, nearly encircling the face, in other species. The facial disk also acts to direct sound into the ears, and a downward-facing, triangular beak minimizes sound reflection away from the face. The shape of the facial disk is adjustable at will to focus sounds more effectively. The prominences above a great horned owl's head are commonly mistaken as its ears. This is not the case; they are merely feather tufts. The ears are on the sides of the head in the usual location.

Talons

While the auditory and visual capabilities of the owl allow it to locate and pursue its prey, the talons, and beak of the owl do the final work. The owl kills its prey using these talons to crush the skull and knead the body. The crushing power of an owl's talons varies according to prey size and type, and by the size of the owl. The burrowing owl (Athene cunicularia), a small, partly insectivorous owl, has a release force of only 5 N. The larger barn owl (Tyto alba) needs a force of 30 N to release its prey, and one of the largest owls, the great horned owl (Bubo virginianus) needs a force over 130 N to release prey in its talons. An owl's talons, like those of most birds of prey, can seem massive in comparison to the body size outside of flight.

The Tasmanian masked owl has some of the proportionally longest talons of any bird of prey; they appear enormous in comparison to the body when fully extended to grasp prey. Owl's claws are sharp and curved. The family Tytonidae has inner and central toes of about equal length, while the family Strigidae has an inner toe that is distinctly shorter than the central one. These different morphologies allow efficiency in capturing prey specific to the different environments they inhabit.

Beak

The beak of the owl is short, curved, and downward facing, and typically hooked at the tip for gripping and tearing its prey. Once prey is captured, the scissor motion of the top and lower bill is used to tear the tissue and kill. The sharp lower edge of the upper bill works in coordination with the sharp upper edge of the lower bill to deliver this motion. The downward-facing beak allows the owl's field of vision to be clear, as well as directing sound into the ears without deflecting sound waves away from the face.

Camouflage

The coloration of the owl's plumage plays a key role in its ability to sit still and blend into the environment, making it nearly invisible to prey. Owls tend to mimic the coloration and sometimes the texture patterns of their surroundings, the barn owl being an exception. Nyctea scandiaca, or the snowy owl, appears nearly bleach-white in color with a few flecks of black, mimicking their snowy surroundings perfectly, while the speckled brown plumage of the Tawny owl (Strix aluco) allows it to lie in wait among the deciduous woodland it prefers for its habitat. Likewise, the mottled wood-owl (Strix ocellata) displays shades of brown, tan, and black, making the owl nearly invisible in the surrounding trees, especially from behind. Usually, the only tell-tale sign of a perched owl is its vocalizations or its vividly colored eyes.

Behavior

Most owls are nocturnal, actively hunting their prey in darkness. Several types of owls, however, are crepuscular - active during the twilight hours of dawn and dusk; one example is the pygmy owl (Glaucidium sp.). A few owls are active during the day, also; examples are the burrowing owl (Speotyto cunicularia) and the short-eared owl (Asio flammeus). Much of the owls' hunting strategy depends on stealth and surprise. Owls have at least two adaptations that aid them in achieving stealth. First, the dull coloration of their feathers can render them almost invisible under certain conditions. Secondly, serrated edges on the leading edge of owls' remiges muffle an owl's wing beats, allowing an owl's flight to be practically silent. Some fish-eating owls, for which silence has no evolutionary advantage, lack this adaptation. An owl's sharp beak and powerful talons allow it to kill its prey before swallowing it completely (if it is not too big). Scientists studying the diets of owls are helped by their habit of regurgitating the indigestible parts of their prey (such as bones, scales, and fur) in the form of pellets. These "owl pellets" are plentiful and easy to interpret, and are often sold by companies to schools for dissection by students as a lesson in biology and ecology.

Breeding and Reproduction

Owl eggs typically have a white color and an almost spherical shape, and range in number from a few to a dozen, depending on species and the particular season; for most, three or four is the more common number. In at least one species, female owls do not mate with the same male for a lifetime. Female burrowing owls commonly travel and find other mates, while the male stays in his territory and mates with other females.

Evolution and Systematics

The systematic placement of owls is disputed. For example, the Sibley - Ahlquist taxonomy of birds finds that, based on DNA-DNA hybridization, owls are more closely related to the nightjars and their allies (Caprimulgiformes) than to the diurnal predators in the order Falconiformes; consequently, the Caprimulgiformes are placed in the Strigiformes, and the owls in general become a family, the Strigidae. A recent study indicates that the drastic rearrangement of the genome of the accipitrids may have obscured any close relationship of theirs with groups such as the owls. In any case, the relationships of the Caprimulgiformes, the owls, the falcons, and the accipitrid raptors are not resolved to satisfaction; currently, a trend to consider each group (with the possible exception of the accipitrids) as a distinct order is increasing.

Some 220 to 225 extant species of owls are known, subdivided into two families: 1. Typical owls or True owl family (Strigidae) and 2. Barn owls family (Tytonidae). Some extinct families have also been erected based on fossil remains; these differ much from modern owls in being less specialized or specialized in a very different way. The Paleocene genera Berruornis and Ogygoptynx show that owls were already present as a distinct lineage some 60-57 million years ago, hence, possibly also some 5 million years earlier, at the extinction of the non-avian dinosaurs. This makes them one of the oldest known groups of non-Galloanserae land birds.

During the Paleocene, other groups of birds, now mostly fill the Strigiformes radiated into ecological niches. The owls as known today, though, evolved their characteristic morphology and adaptations during that time, too. The early Neocene had displaced the other lineages displaced by other bird orders, leaving only barn owls and typical owls. The latter at that time were usually a generic type of (probably earless) owls similar to today's North American spotted owl or the European tawny owl; the diversity in size and ecology found in typical owls today developed only subsequently.

Around the Paleocene-Neocene boundary, barn owls were the dominant group of owls in southern Europe and adjacent Asia at least; the distribution of fossil and present-day owl lineages indicates that their decline is contemporary with the evolution of the different major lineages of typical owls, which for the most part seems to have taken place in Eurasia. In the Americas, rather an expansion of immigrant lineages of ancestral typical owls occurred.

The supposed fossil herons Ardea perplexa (Middle Miocene of Sansan, France) and Ardea lignitum (Late

Pliocene of Germany) were more probably owls; the latter was apparently close to the modern genus Bubo. Judging from this, the Late Miocene remains from France described, as Ardea aureliensis should also be restudied. The Messelasturidae, some of which were believed initially to be basal Strigiformes, are now generally accepted to be diurnal birds of prey showing some convergent evolution towards owls. The taxa often united under Strigogyps were placed formerly in part with the owls, specifically the Sophiornithidae; they appear to be Ameghinornithidae instead.

Family Strigidae (Typical Owls or True Owls)

Genus Aegolius: Saw-whet owls (4 species)

Genus Asio: Eared owls (6-7 species)

Genus Athene: 2-4 species

Genus Bubo: Horned owls, eagle-owls and fish-owls; paraphyletic with Nyctea, Ketupa, and Scotopelia (25 species)

Genus Ciccaba: 4 species

Genus Glaucidium: Pygmy-owls (about 30–35 species)

Genus Grallistrix: Stilt-owls (4 species, prehistoric)

Genus Gymnoglaux: Bare-legged owl or Cuban screech-owl

Genus Lophostrix: Crested owl

Genus Jubula: Maned owl

Genus Megascops: Screech-owls (some 20 species)

Genus Micrathene: Elf owl

Genus Ninox: Australasian hawk-owls (some 20 species)

Genus Nesasio: Fearful owl

Genus Otus: Scops owls (about 45 species)

Genus Pseudoscops: Jamaican owl and possibly striped owl

Genus Ptilopsis: White-faced owls (2 species)

Genus Pulsatrix: Spectacled owls (3 species)

Genus Pyrroglaux: Palau owl

Genus Strix: Earless owls (about 15 species)

Genus Surnia: Northern hawk-owl

Genus Uroglaux: Papuan hawk-owl

Genus Xenoglaux: Long-whiskered owlet

Holistic Information about Owls

Owls are a worldwide order of birds known as Strigiformes and the 216 species range in size from the tiny, sparrow size, Elf Owl to the Eurasian Eagle Owl, which has a wingspan of nearly 2 meters and can weigh almost 5 kg. They are a group of predatory birds, characterized by large forward facing eyes surrounded by a facial disk of short stiff feathers and an upright posture. A large proportion of owls are nocturnal. They occupy an equivalent niche to the diurnal birds of prey such as hawks, falcons, eagles, and buzzards but they are not actually related. The resemblance to the daytime birds of prey is an example of convergent evolution, where both groups have independently evolved several features such as the hooked beak and talons. Owls are all very closely related to each other, much more so than, for instance, the diurnal raptors which include birds as dissimilar as vultures, secretary birds, falcons etc. Even so, owls are separated into two distinct families.

The first family is the Tytonidae, which is made up of 17 species of barn and grass owl, and one species of Bay Owl. Members of this family are quite distinct from other owls and possess several differences. The most obvious external ones being the heart-shaped rather than round facial disk, the longer skull and beak, longer legs, longer and more pointed wings and a forked tail. Grass Owls come from

Africa, South East Asia and Australia and are very similar to Barn Owls but have longer legs.

All of the other 198 owls are in the family Strigidae: The collective noun for owls is a "Parliament".

Food and Hunting

All owls are predators and the size of the prey is reflected generally by the size of the owl. The Elf and Pygmy Owls prey on a variety of insects, spiders and other invertebrates, although, some are quite voracious and can take birds as large as themselves. Eagle Owls, at the other extreme have been known to take such formidable prey as Golden Eagles and a Roe Deer of 13 kg as well as foxes, herons, domestic dogs and there is apparently a report of a large Siberian Eagle Owl taking a ¾ grown wolf !

Hunting, normally, is performed in two different ways. The first is perch hunting, where the bird literally sits and waits on a suitable perch until a prey item is located. The other technique is flight hunting where the owl slowly quarters the ground from a low altitude, looking and listening for prey and diving down when food is detected. The length of the wings is normally a good indication of the preferred method of hunting - short wings for perch hunting and long wings for flight hunting.

Most owls are opportunists and virtually anything that moves is fair game. Some species are more specialized feeders; particularly, the Fish and Fishing Owls. These are the two genera of owls; one from Asia, and the other from Africa, whose members feed mainly on fish, amphibians, and aquatic invertebrates, snatched from beneath the water surface. No other species are quite so specialized but some prefer particular prey. The Milky Eagle Owl of Africa feeds on a wide variety of prey. However, where the ranges overlap, it seems particularly partial to hedgehogs. The Eurasian Eagle Owl would appear to have a vendetta against other birds of prey, especially other owls. In some areas, they form 10% of the bird's diet, which is much greater than one would expect from chance. We must point out the majority of their food is Rabbit.

Pellets

Much is known about the dining habits of owls because of certain inefficiencies of their digestive process. More than 300 species of bird in several different orders are known to regurgitate pellets of indigestible material. This figure includes all owl species. Owl pellets are so informative for several reasons. Firstly owls have comparatively weak bills and often prey that is not too large is swallowed completely, which leaves the skeleton of the prey, including the skull, intact. Unlike most other birds, owls have no crop, and the food passes straight into the foregut (they do not possess a true stomach). The acid in the owl's gut is rather weak with a pH of 2.2-2.5, which is the same as vinegar, this compares to diurnal birds of prey, which have a pH of 1.3-1.8, which is approaching the pH of concentrated hydrochloric acid. This means that owls can only digest the soft tissues. The bones, fur and feathers remain virtually intact. The opening from the foregut into the rest of the digestive tract is small and prevents any undigested material from passing through. Instead, it remains behind where it is compacted into an oval pellet and is then actively regurgitated back up through the esophagus. Pellets, therefore, contain bones including intact skulls, fur, feathers, the chitinous exoskeletons of insects and even the chaetae (bristles) from earthworms and so discovering what owls have been eating is quite straightforward.

Regurgitating a pellet is a voluntary act on the owl's part and in the wild most birds will produce one prior to leaving the roost for hunting. Often, another smaller pellet will be produced during the night before the second main period of hunting around dawn. The size, shape, and appearance of the pellet is normally characteristic of the owl species, certainly amongst British Owls. One interesting fact relating to owl pellets is that the clothes moth, whose larvae sometimes attacks your best woolen suit hanging in the wardrobe, actually evolved from a moth whose larvae feed on the fur and feathers in old owl pellets.

Eyes

Because of their predominantly nocturnal tendencies, owls have evolved several physical adaptations, which facilitate catching prey in the dark. All owls have large forward facing eyes giving good stereoscopic vision, vital for judging distances. Indeed, owls have the most forward facing eyes and hence the best stereoscopic vision of all birds. In smaller species, the head often appears flattened so that the eyes can be as widely spaced as possible to increase the stereoscopic effect. This is often further enhanced by bobbing or weaving the head to give a differing perspective known as the parallax effect.

The eyes are very large, those of a Snowy Owl weighing as much as our own. They are modified in nocturnal species to improve sensitivity in low light intensities. They are tubular, rather than round, giving a relatively large cornea in proportion to the overall size of the eye and enabling more light to enter the eye. The light passes through the pupil (which can be closed by the iris to a small pinprick in bright light or opened so wide that virtually no iris is visible at night) to the lens. This is large and convex, causing the image to be focused nearer to the lens hence retaining maximum brightness. One drawback is that owls are long-sighted, and cannot focus on objects, which are too close. Tactile bristles around the beak partially compensate for this. The tubular shape also gives a comparatively large retina size which is packed full of light sensitive rods, about 56,000 per square mm. in the Tawny Owl. These rods are far more sensitive than cones at low light levels. The phenomenal light gathering properties of the owls eye is further enhanced in many species by a reflective layer behind the retina, called the tapetum lucidum, which reflects back onto the rods any light that may have passed through the retina without hitting one the first time. Tawny owls would appear to have the best developed eyes of all the owls, indeed of all vertebrates, being probably about 100 times more sensitive at low light levels than our own.

As well as rods, all owls possess color sensitive cones in their eyes. Although having fewer light sensitive cones than humans, they can probably detect colors to some extent. They are certainly not blind in daylight and some, like the Eagle Owl, have better daytime vision than us. Our nighttime vision, however, is better than some diurnal

Pygmy Owls. Owls are unable to move their eyes in the sockets because of the size and tubular shape. To compensate, they have a deceptively long flexible neck, which enables them to turn their head 2700 in either direction horizontally and at least 900 vertically.

Plumage

The plumage of owls is adapted in several ways. Being mainly nocturnal, owls have no need to display visually and so their markings are normally mottled shades of black, brown, and grey, patterned to conceal them during the daytime from possible predators and other birds, which might mob them. The snowy owl, although distinctive with its white plumage, blends perfectly with the arctic snow which is often still on the ground when the female is incubating her eggs. The Barn Owl's plumage may appear more puzzling. If viewed from above, the most likely direction from attack whilst hunting over its favorite habitat of long rank grassland with its brown dead stems of long grass, it almost disappears from sight as it does in its original nesting sites in sandstone cliffs. The white under-parts may make it less visible against the sky when it hunts in daylight, which it does when food is in short supply or when there are chicks in the nest. The ear tufts possessed by some species are just feathers and nothing to do with the ears. They can be used as camouflage, by breaking up the owls outline and their posture communicates their mood.

Most owls have a defensive posture, which is used to intimidate potential enemies. It varies from species to species but usually involves the wings being spread out around the body with the head lowered down and the feathers fluffed up. This all gives the impression that the owl is much bigger and when accompanied by "bill clicking", hissing and swaying the body from side to side is normally enough to drive off most predators. The feathers on owls' wings have several unique features. Firstly, they have a downy upper surface, which reduces the noise as they move over each other as the wing beats. Secondly, the outer primaries have a stiff comb like fringe on their leading edge, which modifies the airflow over them as they cut through the air, which reduces noise. Finally, on the trailing edge of the wing, the feathers have a soft hair like fringe, which reduces air turbulence behind the wing. The net result of these modifications is totally silent flight even at ultrasonic level which means that prey are not forewarned of attack and the bird can use its amazing hearing at its best without the noisy wing beats possessed by other birds.

In addition, the plumage of owls is very thick and downy, retaining vital body heat when hunting at night. This makes owls look allot larger than they actually are. Most owls are deceptively light for their size; also most have quite large and broad wings, which with the low body weight gives them a very low wing loading, which is why their flight is so buoyant and moth like. This allows them to fly at a very low speed, which improves the chances of locating prey, and it lets them carry quite heavy loads back to the nest or roost. The thick soft plumage insulates the owl whilst it is exposed to cold nighttime temperatures, and in most species (certainly all northerly species), the leg feathers extend right down to the feet and sometimes over the toes as well.

Feet

Owl legs and feet vary considerably between species, depending on its ecology. All species have surprisingly long legs, most of which is hidden in the soft belly feathers. The Barn Owl has long sparsely feathered legs using them to snatch prey from long vegetation. Despite ranging into temperate zones, the minimal feathering probably prevents the bird's legs becoming sodden if it hunts in damp grass. The Grass Owls from Africa, Asia and Australia hunt in even longer grass and their legs are even longer.

The European sub-species of Eagle Owl are found in Northern and Central Europe often at quite high altitudes, and its feet and toes are covered in feathers. The Snowy Owl from around the Arctic Circle has the most heavily covered of all. These feathers also help protect the owl from being bitten by prey. At the other end of the scale, most of the Fish and Fishing Owls have very bare legs and feet, again feathers in this situation would become waterlogged. The Burrowing Owl from North and South America lives on the Prairies and Pampas in the burrows of Prairie Marmots and Viscaschas, occasionally it even does some digging of its own using its very long legs. It is extremely terrestrial, often chasing invertebrates on foot and the extra height gained from its legs gives it a better view of predators on the flat grass plains. All owls have a flexible outer toe. At rest, it points roughly forward sticking out slightly to the side, but it can be angled backwards giving an increased area to catch prey and a better grip.

Breeding

No owl can be said to build its own nest in the proper sense of the word. Burrowing Owls may extend an already existing burrow, and some of the large owls may scratch a slight depression or scrape in loose earth but nothing that compares with the marvels produced by some other birds. Nest sites vary but tree hollows and cavities are the preferred site for many species. The Elf Owl nests in holes in Sanguaro cactuses made by the Gila Woodpecker, the Burrowing Owl as mentioned in animal burrows, Grass owls trample tunnels and nest chambers in long grass, whilst larger species scratch a scrape or use old stick nests of crows, pigeons, or diurnal birds of prey. The American Great Horned Owl often drives out quite large raptors such as red-tailed buzzards and takes over their nests. Tawny Owls often use old squirrel drays, and Barn Owls in Africa use the giant stick structures made by Hammerkops (a stork like bird) occasionally nesting communally. Many species can be encouraged to use nest boxes and they may even be used in preference over natural sites. Many owls are quite territorial, especially during the breeding season, sometimes even other owls species are not tolerated, at the other extreme some species will defend the nest site itself but will share hunting habitat.

Breeding seasons are always geared so that rearing chicks coincides with the peak availability of food; this is not always when prey is most numerous but when it is able to be caught in increased numbers. This may be as a result of decreased vegetation cover, or when the prey are more active or vocal say in defense of their territories allowing them to be caught more easily. In some species this often means that the eggs are laid very early, sometimes with snow still on the ground, this also gives the relatively slow growing young more time to learn how to hunt more efficiently before the following winter.

All owls lay white eggs, which suggest that they all evolved from a hole-nesting ancestor. Elaborate markings to conceal the eggs from predators are not needed in dark holes indeed the whiteness may make them more visible to the parents. The egg above is being candled. A light is shone through the egg so we can see the developing embryo. (The solid Pink Blob in the Middle) The blood vessels are forming to feed the developing chick and to allow Oxygen to be absorbed into the blood stream. The Yolk inside the egg is not visible but is there to feed the baby as it grows.

The number of eggs laid varies from species to species, year to year and between individual birds. In general, larger birds lay fewer eggs, and birds from tropical regions lay fewer than birds from more extreme latitudes. In temperate and sub-arctic regions some species like the Snowy Owl the Short-eared Owl and the Barn Owl can increase the size of the clutch as prey availability increases, in years when the lemming or vole populations crash, breeding may be abandoned totally.

Owl eggs are relatively spherical to a greater or lesser extent. In most species, the female starts incubating as soon as the first egg is laid. The eggs are laid at intervals of at least a day, often more, resulting in what is known as asynchronies hatching, where the eldest chick can be up to 2 weeks older than the youngest. The means that each chick reaches its peak food demand period at different times and so spreading the load. In lean years the oldest and therefore strongest chicks at least will survive, the younger chicks, once dead, may even be utilized to feed the others.

During incubation, and until the smallest chick is large enough to maintain its own body temperature, all the food is provided by the male, the female rarely leaving the nest site. She dispatches the food and feeds the chicks small slivers until they can swallow the prey whole, she then helps the male with the hunting. The age of fledging varies greatly and some species even remain in the area until the following year. Gradually the young learn to hunt, often starting on insects or food brought in by the parents, who may still be alive. Most species are independent by their first winter and in many cases, their parents drive the young actively away prior to this.

Calls

Like most birds, owls sing to attract a mate and defend their territories although perhaps not as musical as the songs of thrushes and warblers, the owl song functions in exactly the same way. The noises made by owls include the melodic trills of pygmy owls, the bell like chimes of the European Scops Owl, the classic hoot of the Tawny Owl (and it's not so musical shriek), the strangulated screech of the Barn Owl and the deep resonant hoots of Eagle Owls.

In general, the pitch and volume of the call reflect the size of the territory of the owl. Of course, there are exceptions. Many forest species have loud calls for the size of their territories but this is because of the sound-deadening effect of the trees. The Flammulated Owl from the western United States is a similar size to the Western Screech Owl, which inhabits the same part of the world, but its territory is much larger and hence it has a loud deep call, more becoming of a much larger bird. Diurnal species are far less vocal than nocturnal ones, being able to use visual displays to advertise for partners and defend their territories.

Don't forget, Owls don't breed very well when it's raining because it's too wet to Woo.....!

Symbolisms about Owls

Throughout human history, owls have variously symbolized dread, knowledge, wisdom, death, and religious beliefs in a spirit world. In most Western cultures, views of owls have changed drastically over time. Owls can serve simultaneously as indicators of scarce native habitats and of local cultural and religious beliefs. Understanding historical and current ways in which owls are viewed, and not imposing Western views on other cultures, is an important and necessary context for crafting owl conservation approaches palatable to local peoples.

I am a brother to dragons, and a companion to owls. Job 30: 29

Thou makest darkness, and it is night: wherein all the beasts of the forest do creep forth. Psalms 104:20

Long before there were ornithologists and graduate students, keen observers in other tribes and bands roamed the forests and plains. In their search for resources, they encountered winged denizens of the night and incorporated such spectral figures into their lore and culture.

The North American Cherokees call them uguuk, the Russians sovah, the Mexicans tecelote, the Ecuadorians huhua or lechusas, and aboriginal peoples of the Kaurna area of Australia winta. For centuries, indeed millennia, owls have played diverse and fascinating roles in a wide array of myths and legends. In this era of rapidly shrinking habitats for many owls of the world, a first step toward garnering concern for their conservation is to better understand of the role of owls in cultural stories and lore. Here, one has a hope to foster an appreciation for the breadth by which owls have been invited into the mythos of human societies. One offers this as a celebration of the diversity of response to owls by humankind the world over. The spectrum of human responses is both remarkable and wonderful. No other bird family has aroused more universal fascination and interest, and better can serve as a basis for conservation. For conservation, we must proceed from respect for diverse cultures and creeds, as much as for the organisms that share our sphere.

African Culture

Among the Kikuyu of Kenya, it was believed that owls were harbingers of death. If one saw an owl or heard its hoot, someone was going to die. In general, owls are viewed as harbingers of bad luck, ill health, or death. The belief is widespread even today.

Ancient European and Modern Western Culture

The modern West generally associates owls with wisdom and vigilance. This link goes back at least as far as Ancient Greece, where Athens, noted for art and scholarship, and Athena, Athens' patron goddess and the goddess of wisdom, had the owl as a symbol. Marija Gimbutas traces veneration of the owl as a goddess, among other birds, to the culture of Old Europe, long pre-dating Indo-European cultures.

T.F. Thiselton-Dyer, in his 1883 'Folklore of Shakespeare', says that "from the earliest period it has been considered a bird of ill-omen", and Pliny tells us how, on one occasion, even Rome itself underwent a lustration, because one of them strayed into the Capitol. He represents it also as a funereal bird, a monster of the night, the very abomination of human kind. Virgil describes its death-howl from the top of the temple by night, a circumstance introduced as a precursor of Dido's death. Ovid, too, constantly speaks of this bird's presence as an evil omen; and indeed the same notions respecting it may be found among the writings of most of the ancient poets. A list of "omens drear" in John Keats' Hyperion includes the "gloom-bird's hated screech". Pliny the Elder reports that owl's eggs were used commonly as a hangover cure.

Asia

In Mongolia, the owl is regarded as a benign omen. In one story, Genghis Khan was hiding from enemies in a small coppice when an owl roosted in the tree above him, which caused his pursuers to think no man could be hidden there. In modern Japan, owls are regarded as lucky and are carried in the form of a talisman or charm.

Hinduism

In Hinduism, an owl is the vahana, the mount, of the Goddess Lakshmi.

Native American Culture

People often allude to the reputation of owls as bearers of supernatural danger when they tell misbehaving children, "The owls will get you" and in most Native American folklore, owls are a symbol of death. According to Apache and Seminole tribes, hearing owls hooting is considered the subject of numerous "bogeyman" stories told to warn children to remain indoors at night or not cry too much, otherwise the owl may carry them away. In some tribal legends, owls are associated with spirits of the dead, and the bony circles around an owl's eyes are said to comprise the fingernails of apparitional humans. Sometimes owls are said to carry messages from beyond the grave or deliver supernatural warnings to people who have broken tribal taboos.

The Aztecs and Maya, along with other natives of Mesoamerica, considered the owl a symbol of death and destruction. In fact, the Aztec god of death, Mictlantecuhtli, was often depicted with owls. There is an old saying in Mexico that is still in use "When the owl cries or sings, the Indian dies". The Popol Vuh, a Mayan religious text, describes owls as messengers of Xibalba (the Mayan "Place of Fright").

The belief that owls are messengers and harbingers of the dark powers is also found among the Hocągara (Winnebago) of Wisconsin. When in earlier days the Hocągara committed the sin of killing enemies while they were within the sanctuary of the chief's lodge, an owl appeared and spoke to them in the voice of a human, saying, "From now on the Hocągara will have no luck". This marked the beginning of the decline of their tribe. An owl appeared to Glory of the Morning, the only female chief of the Hocąk nation, and uttered her name. Soon afterwards she died.

According to the culture of the Hopi, an Uto-Aztec tribe, taboos surround owls, which are associated with sorcery and other evils. Ojibwe tribes, as well as their Aboriginal Canadian counterparts, used an owl as a symbol for both evil and death. In addition, they used owls as a symbol of very high status of spiritual leaders of their spirituality. Pawnee tribes viewed owls as the symbol of protection from any danger within their realms. Pueblo people associated owls with Skeleton Man, the god of death and spirit of fertility. Yakama tribes use an owl as a powerful totem, often to guide where and how forests and natural resources are useful with management.

Motifs and Mythology

One can see a likeness between the old, animist forest, where one could not be sure whether a screech owl's call came from a bird or an Omah, and the evolutionary forest, with its unclear distinctions between tree and fungus, flower and fir cone. The tree-fungus relationship is as mysterious in its origins and implications as the owl-Omah one. Both belong to a world that goes deeper than appearances, where a buried interconnectedness of phenomena renders behavior ambiguous, where one cannot walk a straight line.

Owls have always been part of the root metaphors of how humans relate to the land. One of the earliest human drawings dating back to the early Paleolithic period was of a family of Snowy Owls (Nyctea scandiaca) painted on a cave wall in France. Rock paintings or petroglyphs of owls have been found in other disparate locations including the Victoria River region of northern Australia and the lower Columbia River area of Washington State, USA. Owls played roles in the ancient Mayan cultures of Mesoamerica. A carved bas-relief of the ancient Mayan Ruler-III of Dos Pilas, in what is now Guatemala, following the death of Ruler-II in 726 C.E., is shown adorned with a screech owl, apparently a symbol of ruling power or the resurrection of government.

Owls also are very much a part of modern culture, in the sky as well as on the land. In the constellation Ursa Major, at a most dim magnitude of 11.20, is an irregular planetary nebula designated by astronomers as the Owl Nebula (more formally called M97 or NGC3587). In a more terrestrial venue, a query of the U.S. Geological Survey database on place names revealed 576 features in the United States in some way named "owl," such as Owlshead Canyon, Owl Mine, Owl Creek, and Owl Hollow. Records of the Canadian Permanent Committee on Geographical Names lists 88 current and 17 additional historic places in some way named "owl". Doubtless, many other countries have similar designations.

Etymologically, the word "owl" goes back to the Middle English word "oule", which may derive from the Old English "ille", which is cognate with the Low German "ule", in turn going back to the German "eule". The ultimate root of the modern word "owl" was presumed to be a proto-Germanic word "uwwalo" or possibly "uwwilo". Another derivation of "owl" is the Icelandic "ugla," which is cognate with "uggligr," which gave rise to the Scandinavian "ugly," which led to the Middle English "ugly" and the Modern English word "ugly." The Icelandic "uggligr" does not mean "ugly" in modern connotations (that is, unpleasant to behold), but rather it means "fearful or dreadful." This is precisely the connotation of owl symbols and totems in many myths and legends. Thus, the very names that we use often speak of a deep history of traditional viewpoints and cultural perspectives. Further, in Hindi, owl is "ul" (similar to the German "eule" or Low German "ule") or "ulu" if referring to one of the large owls (the Hindi or Urdu term for smaller owls is "coscoot").

The ancient Roman "bubo," the ancient Greek "buas," the modern Hindi "ulu," and the Modern Hebrew "o-ah" are obvious onomatopoeias, as is the modern Nepali "huhu."

An awareness and understanding of the deep, complex perceptions of owls in the past may help support efforts to protect those species today. For example, the ancient cultural importance of owls in Europe helps modern conservationists there. The same is true in America. The blend of traditions carried to the U.S. by white immigrants and black slaves from West Africa means that North American owl species have a strong cultural profile that may aid conservation measures. There were so many beliefs about owls amongst African-American slave and ex-slave communities in pre-historical period. Such beliefs seeped into the dominant European-American culture just in the way that African rhythms were given to the world through blues and jazz music of black North America. Thus, current U.S. folklore about owls is an eclectic blend of European and African traditions, and Native American and Asian as well.

This can be extended to an environmental principle for the West (meaning all areas occupied by those of European descent, and by mixed-race societies such as South Africa, where a highly developed conservation tradition exists.). Any animal or flower with a strong cultural profile, no matter how negative that cultural perception may once have been (such as with bats, wolves, sharks, and owls) is at a major advantage, for conservation, over an animal with no cultural profile whatsoever (such some rodents and sparrows). The advantage is that they are rooted and recognized in the social consciousness. In the case of owls, the deep fears and anxieties they generated and the prophetic status they once held (and still hold) present environmentalists with a handle with which to engage the interest and sympathies of a wider audience. However, the critical element in these situations is the fact that most Western cultures no longer perceive owls as omens of evil, or retain only the dimmest vestiges of these old beliefs.

A good analogy would be our celebration of witches and the like at Halloween. We can enjoy the witches' Sabbath today precisely because we no longer believe in demonic power or demonic possession. It has been degraded culturally to the status of a parody. In a sense, we mock our former frailties when we dress up as witches and ghouls. But we could never have done that without being detached from the intrinsic and original meaning of the event. In the 17th century, witches and Halloween were literally deadly serious. Now they are entertainment.

However, for some or even many Africans, Native North Americans, Asians, and South Americans, these perceptions of owls are living traditions with deep and powerful roots. For example, in Africa, owls are believed still genuinely to be evil. Heimo Mikkola's surveys of attitudes towards owls in Malawi revealed that owls were regarded as bad birds by a very high percentage (~80%) of the people surveyed. Most people do not like owls and regard them as evil in West Africa. The standard pigeon English name for owl in West Africa is "witchbird". Rather than garnering support for endangered species such as the Congo Bay Owl (Phodilus prigoginei), the ancient African mythic traditions relating to owls may present a barrier to their conservation. A classic parallel case is the Aye-aye

(Daubentonia madagascariensis) of Madagascar where the beast, down to a last few dozen, has been persecuted ruthlessly because of its cultural profile as a witch-creature. The challenge for conservationists is to turn the barrier to an advantage by understanding the cultural moors and helping to craft conservation actions taking these into account.

Conservationists should understand the role a bird like an owl may play in some societies. Conservation policies for a Red Data species, such as the Congo Bay Owl, should not be formulated without understanding local attitudes and any uses of that particular species. Conservationists too often inculcate their own positive view of the animal in question, but fail to change local cultural attitudes, that is, to replace deep fear with admiration and respect. The environmental community cannot tackle owl conservation without understanding the cultural profiles which most owl species have had foist upon them, some positive, and many negative.

Markers of Gods, Knowledge, Wisdom, and Fertility

Many children have grown up with nursery stories of wise old owls. From the ancient Greek legends to the wise owls in Wini the Pooh and The Owl and The Pussycat, we have all seen images in folk tales of owls as the quintessential bearers of knowledge and sagacity. From ancient Athens, the silver four-drachma coin bore the image of the owl on the obverse side as a symbol of the city's patron, Athene pronoia, the Greek goddess of wisdom who, in an earlier incarnation, was goddess of darkness. The owl, whose modern scientific name Athene carries this heritage came to represent wisdom from its association with the dark. The owl was also the guardian of the Acropolis, and the Roman statesman Pliny the Elder wrote that owls foretell only evil and is to be dreaded more than all other birds.

In many other cultures, owls represent wisdom and knowledge because their nocturnal vigilance is associated with that of the studious scholar or wise elder. According to one Christian tradition, owls represent the wisdom of Christ, which appeared amid the darkness of the unconverted. To early Christian Gnostics, the owl is associated with Lilith, the first wife of Adam who refused his advances and control. The owl had a place as a symbol in the King Arthurian legends since the sorcerer Merlin was depicted always with an owl on his shoulder. In Japan, owl pictures and figurines have been placed in homes to ward off famine or epidemics.

Some Native American cultures link owls with supernatural knowledge and divination. In the Menominee myth of The Origin of Night and Day, Wapus (rabbit) encounters Totoba (the saw-whet owl, Aegolious acadicus) and the two battle for daylight (wabon) and darkness (unitipaqkot) by repeating those words. Totoba errs and repeats "wabon" and daylight wins, but Wapus permits that night should also have a chance for the benefit of the conquered, and thus day and night were born. The Pawnees view the owl as a symbol of protection; the Ojibwa, a symbol of evil and death, as well as a symbol of very high status of spiritual leaders of their religion; and the Pueblo, associated with Skeleton Man, the god of death and spirit of fertility. On a warm afternoon in August 1985, one researcher observed Ojibwa peoples at a weekend cultural celebration in Duluth, Minnesota using dried wings of Great Horned Owls (Bubo virginianas) as hand-held fans to cool themselves after participating in native dances.

In the book "Mother Earth Spirituality", the four directions of the Sacred Hoop (the four quarters with the power of earth and sky and all related life) of Native Americans were mentioned. In this description, the Snowy Owl represents the North and the north wind. The traditional Oglala Sioux Indians (from central North America) admired the Snowy Owl, and warriors who had excelled in combat were allowed to wear a cap of owl feathers to signify their bravery. An old-time society of the Sioux was called The Owl Lodge. This society believed that nature forces would favor those who wore owl feathers and, as a result, their vision would become increased. The owl is a good example of a creature that possesses special powers not found in other animals.

Some Native American nations in the U.S. have strong taboos against owls. For example, the Apaches view the owl as the most feared of all creatures. In 1997, a spokesperson for the Apache told about the deep taboo against owls. Historically, Apaches shared the widespread Athabascan fears of owls as the embodied spirit of Apache dead. John Bourke, in his "Apache Campaign in the Sierra Madre", related a famous story of how Apache scouts tracking Geronimo became terrified when one of the U.S. soldiers found and brought along a Great Horned Owl. The scouts told Bourke that it was a bird of ill omen and that they could not hope to capture the Chiricahua renegades if they took the bird with them. The soldier had to leave the owl behind. In another example, the consortium of Yakama tribes in Washington State in the U.S. uses the owl as a powerful totem. Such taboos or totems often guide where and how forests and natural resources are used and managed, even to this day and even with the proliferation of "scientific" forestry on Native American lands.

The Blakiston's Fish Owl (Ketupa blakistoni) was called "Kotan Kor Kamuv" (God of the Village) by the Ainu, the native peoples of Hokkaido, Japan. The traditional Ainu people were hunter-gatherers and believed that all animals were divine; most admired were bear and the fish owl. The owls were held in a particular esteem, associated with fish (salmonids) and lived in many of the same riverside locations. The Fish Owl Ceremony, which returned the spirit of fish owls to the god's world, was conducted until the 1930's.

Owls have played various roles in Russian traditions. For example, in Slavonic cultures, owls were believed to announce deaths and disasters. Russians and Ukrainians sometimes call an unfriendly person a "sych", which is also the Russian common name of the Little Owl (Domovoy Sych; Athene noctua). Traditionally, little owls have been disliked; feared by the people, who believe that these birds announce the death. However, Russian common names of other owls, such as the Scops Owl (Otus scops) - Splyushka, resembling its call, or Zorka, meaning dawn - do not carry this negative connotation. In old Armenian tales, owls were associated with the devil. In Central Asia, feathers of the Northern Eagle Owl (Bubo bubo), particularly from its breast and belly, were valued as precious amulets protecting children and livestock from evil spirits. Talons of the Northern Eagle Owl were said to ward off diseases and cure infertility in women. In some stories, the Klamath Mountain Indians of northwestern California, USA, referred to "Bigfoot", an elusive bipedal hominid supposedly inhabiting the deep forests, as Omah. In Russian literature, Central Asia is the region to the east of the Caspian Sea including Turkmenistan, Uzbekistan, Tajikistan, and Kyrgyzstan.

Owls in Lore and Culture

In many cultures, owls signal an underworld or serve to represent human spirits after death; in other cultures, owls represent supportive spirit helpers and allow humans (often shamans) to connect with or utilize their supernatural powers. Among some native groups in the Pacific Northwest of USA, owls served to bring shamans in contact with the dead, provided power for seeing at night, or gave power that enabled a shaman to find lost objects. As with the owls of the ancient Roman statesman Pliny the Elder, many forest owls have played key roles as signalers of death. The mountain tribes of Myanmar (Burma) know the plaintive song of the Mountain Scops Owl (Otus spilocephalus) in such legends. In one Navajo myth, after death the soul assumes the form of an owl.

The aboriginal peoples of North Queensland, Australia, view owls in a similar way. In January 2000, a female aboriginal elder relayed that owls are special to her people. A little apologetically, she added that owls are also considered an ill omen, signifying a death in the family - but only if the owl hung around the home site for several days.

In China, owlets have been believed to pluck out their mothers' eyes. The owl's night excursions, staring eyes, and strange call have led to a widespread association with occult powers. The bird's superb night vision may underlie its connection with prophecy, and the reputation for being all-seeing could arise from its ability to turn its head through almost 180 degrees. In a similar vein, on Andros Island, Bahamas, an historically extinct species of flightless owl, Tyto pollens, scientifically known only from sub-fossils, stood one meter tall and may have been the source of old local legends of "chickcharnies" or aggressive leprachaun-like imps that wreak havoc, have three toes, and can turn their heads all the way around. This owl likely inhabited the dense stands of old-growth Caribbean pine (Pinus caribbeanensis), so much of which had been clear-cut on Andros during the latter 20th century by American companies.

Along the northwest coast of Alaska, the Yup'ik peoples made masks for a final winter ceremony called the Agayuyaraq ("way, or process, of requesting"), also referred to as Kelek ("Inviting-in Feast") or the Masquerade. This complex ceremony involved singing songs of supplication to the animals' yuit ("their persons"), accompanied by the performance of masked dances, under the direction of the shaman. In preparation for the ceremony, the shaman directed the construction of the masks, through which the spirits revealed themselves as simultaneously dangerous and helpful. The helping spirits often took the form of an owl. The majority of masks contained feathers from snowy owls. Carvers strove to represent the helping spirits or animal yuit they had encountered in a vision, dream, or experience. In all cases, the wearer was infused with the spirit of the creature represented. Together with other events, the ceremony embodied a cyclical view of the universe whereby right action in the past and present reproduced abundance in the future.

On Java and Borneo, the Collared Scops Owl (Otus bakkamoena) has survived thanks in part to the fact that it is viewed in legends there with reverence or as an ill omen. These owls are taken in China and Korea for medicinal use and many have been lost annually for such purposes. Shakespeare wrote of "The owl, night's herald" and recognized the role that owls have as the "fatal bellmen" of that final deepest sleep. In this way, owls have been seen as harbingers of eschatology or the ultimate fate of humans.

Throughout India, owls are construed as bad omens, messengers of ill luck, or servants of the dead. In general, owls often have been treated badly both in daily life and even in Indian literature. For example, in India it is very common to call a foolish person "an owl". Nevertheless, in Indian mythology the owl has been treated at times reverently and given some place of prestige. For instance, Laxmi, the Hindu goddess of money and wealth, rides on an owl. Even in present times, some people of India, particularly Bengali, believe that if a white owl enters a home it is treated as a good omen by relating it to the possible flow of wealth or money into that home.

In India, the Forest Eagle owl is known to take peafowl, jungle fowl, hares, jackals, and even young barking deer. A renowned Indian ornithologist (Late Dr. Salim Ali) noted that its cry is low, deep, and far-sounding moaning hoot and a blood-curdling shriek as of a woman in grief, earning this creature the name of "Devil Bird". The call of the Ceylon Forest Eagle owl subspecies (Bubo nipalensis blighi) consists of "shrieks such as of a woman being strangled" but that "the dreadful shrieks and strangulating noises are merely its 'mating love-song,' which would also account for their rare and periodic occurrence". In related accounts, Ali described its noises as "a variety of weird, eerie shrieks and chuckles" and a scream "like that of a demented person casting himself over a precipice". Some researchers also noted in history that eagle owls have been variously called as Bird of Evil Omen, Death Owl, Ghost Owl, Mystery Owl, Knows-All Owl, and even Rat Owl.

The Devil Bird or Devil Owl can be found in graveyards and big dead trees in India. This owl species associates with old forests, big old trees, and death. Graveyards often contain the last old growth trees. In India, the Muslims, especially, revere everything in a cemetery including the vegetation. Thus, the eerie cries of the Devil Owl are heard mostly in cemeteries, foretelling the death. Moreover, converge myth, culture, and biology to a consistent whole; they should for successful conservation of cultures, people, and wildlife.

In Indian jungles, the Brown Wood Owl (Strix leptogrammica), Forest Eagle owl (Bubo nipalensis), and Brown Fish Owl (B. zeylonensis) are found in dense riparian forests of Ficus near streams and ponds, sites often considered as sacred groves, or in cemeteries that bear the last of the largest trees with cavities and hollows in an area. Old-forest owls, particularly the Forest Eagle owl, play major roles in many Nepali and Hindu legends.

As heard calling at night from cemeteries and sacred groves, such owls are thought to have captured the spirit of a person departed from this world. In one sense, then, many of these owl species can serve as indicators of the religious value of a forest; conserving the religious site equally conserves key roost or nest sites.

Members of the animistic Garo Hills Tribe of Meghalaya, northeast India, call owls as dopo or petcha. Along with nightjars, they also refer to owls as doang, which means birds that are believed to call out at night when a person is going to die; its cry denotes the death of a person. According to Baban, in northern India and Nepal, small owls, coscoot, are very good. They come to the house and take mice and bad insects. However, large owls, ulu, are very bad. They will come and perch on the house at night, and that is bad, because if someone comes to the house and then says your name, ulu will catch it and have it. It will then wait until all is dark and quiet, and call you with your name, "come out!" and you will come out to see, and there will be no one there. It will call you out again by name, "come out!" Then in 10 days or 20 days, by repeating your name repeatedly, your life will ebb and death will surely follow.

In one of the owl myths, the culprit is the Spotted Owlet (Athene brama), a common resident of city and garden throughout greater India. Instead of catching your name, it will catch a stone you throw at it and slowly grind down the stone. As the stone decays, so does your life. If you should find the owl, it would appear to befriend you, to tell you things. It will be obliged to tell you how to prepare itself for cooking. Little by little, over the nights it will appear on the roof and will teach you how to prepare its claws, its feet, its legs, its back, its wings, and its neck. In its instructions, when it finally reaches its neck, on the 39th day, you must grab it and slice its head off, or you will be dead the next morning.

Further, the tiger, the ulu, and other animals of the jungle that themselves prey on animals is very bad, but the ones that eat only vegetation are very good. This distinction may have arisen in India and Nepal from the undesirable habit of carnivory and that herbivory or insectivory better fit the Hindu vegetarian philosophy.

If a Barred Owlet or Spotted Owlet is heard, then soon a child will fall ill. Such small owls those with similar tooting calls are all referred to as Nattu collectively. The Mottled Wood Owl is called Kalam Kozhi. Kalam means god of death, and Kozhi means chicken or fowl, so the overall name means fowl of death. If heard when you are ill or diseased, it means you will die. Horned owls (Bubo) can land on the head of hunters and the hunter will die of fright.

The Tharu Tribe inhabits riverine grassland environments in northern Uttar Pradesh along the border with Nepal. There, the owls are associated with bad fortune, but it is only the larger owls that bear this stigma. (Large) owls are often associated with the devil, and are not auspicious. The female Eagle Owl, Bubo bubo, can get very annoyed (if harassed) and then she feigns death. You can approach the owls, even turn it over and it will not move. The male owl will keep giving calls and the female will finally respond with a quiet call.

Local shamans can kill an owl and take its soul, its power, and put its power into a tabich (a talisman worn around the neck). Then the owl power will guide you to find wealth or to find people of wealth (it does not give you wealth; it only guides you). If any part of the owl (such as its talons or feathers) is taken for its power, only the shamans know this.

There is a Garo tribe story of an owl: An owl was with a Hill Myna and said, "I shall lay eggs as beautiful as yours." The owl said this same thing day after day but never laid the eggs. Therefore, this led to the (Garo) phrase and concept of saying one thing and doing another. In Kaziranga National Park in Assam, northeast India, a large owl appears in yard and calls Hoo-Doo; it is an "indicator" that something will happen like bad omen or illness. The Asamese word for owl itself is hoodoo, an obvious onomatopoeia.

But if a small owl calls, it signals that a young girl is about to reach puberty and womanhood, which in Assamese culture is cause of celebration and a big party. In far south India (Kerala), Thekkady village on the border of Periyar Tiger Reserve, local people consider owls in a very superstitious manner because owls - particularly Mottled Wood Owls (Strix oscellata) - "have a very eerie kind of a call". The Mottled Wood Owl's call is viewed as "a heralding of the Goddess of Death" and signals, "Death is imminent." Even in the sound tracks of films, owl calls are used to impart a sense of ominous tension when waiting for something grievous to happen. Such ideas are held not just by the local tribe but are transferred among many indigenous communities and thus serve to reinforce and complement such beliefs in this way.

In other parts of India or Asia, smaller owls can bring good luck or fortune. Any owl is associated with any element of goodness in south India, and that owls always herald something unfavorable. The booming call of the Brown Fish Owl (Bubo zeylonensis) in the night, the male will call once and the female will call twice. This has become a kind of saying, a proverb, that when the male owl makes one call, the female calls twice, referring to if a husband and wife of the house fight, and the wife will be more vociferous. Sometimes an owl - likely the smaller owls, such as Asian Barred Owlet (Glaucidium cuculoides) or Jungle Barred Owlet (Glaucidium radiatum) - might come into the house after insects. It is believed that when this happens, the owl will bring illness.

In ancient Egypt, India, China, Japan, and Central and North America, owls were the bird of death. In other cultures and religions, however, such as ancient Greece, they bore the role of supernatural protector. Some Native Americans, for instance, wore owl feathers as magic talismans.

Owls in Tribal Folklores

According to Gavin Robinson, the (white) director of a game and ostrich farm north of Bulawayo ("The Cawston Block") in western Zimbabwe, the local indigenous people there (Shonas) view Ground Hornbills and owls as evil or as portending death. If an owl lands on your house, it is believed that ill luck or illness per se, will follow. This is especially believed of the Common Barn Owl (called "Screech Owl" there) because of its commensal association with humans and houses. The witch doctors take owls and use their talons and beaks for medicines, which help them, harm other people - very powerful medicine.

In Namibia, Balozi tribe in the eastern Caprivi Strip on Impalila Island in the middle of the Zambezi River believes that owls in his tribe are thought to bring disease. When owls enter the village, they are shunned or shot, in part because the larger owls such as Giant Eagle-Owls take chickens, but also because they are thought to induce disease merely by their presence. In Bulawaya, Zimbabwe, Common Barn Owls are seen as a "witch's bird" among the local black population.

In the village of Bongonde Drapeau (between Mbandaka and Bikoro), local tribe Bantu believes that owls are "dreaded". This was essentially repeated by other Bantu people in other villages too. However, it is surprised at how willing local villagers locate owls at night. Apparently, local beliefs did not inhibit at least some of the people in seeking out owls in the forest at night.

In Cameroon, witches are reported to transform into owls. In addition, owls announce bad news or foretell death. If you make a call like an owl, people will scold you to stop, saying, and "Are you a witch?" In Cameroon there are also folk tales about bats which people do not distinguish as a mammal and think of as intermediate between a mammal and a bird (in fact, the French word for bat there is "duplicate").

In the Democratic Republic of the Congo, perception of local people about owls is "not good". When you hear an owl, it means bad news, such as "perhaps a relative has died back in Kinshasa" (the capital city). So owls are feared, as they "bring prophecy." Although owls are known to be bad, they are also occasionally eaten (as are most other animals in the region). Both large and small owls are eaten. Owls are "bad species" by custom, but only against men, not women. There is no formal hunting of owls. Owls are killed when they are seen, though, opportunistically.

African Wood Owl, Strix woodfordii, a species that most commonly found in and around the villages and secondary forests also gained beliefs. The local tribe noted that there are "two kinds of cries" of the (wood) owl: the bad cries from the male, which mean bad news, and another cry from the female, which does not mean bad news. What they referred to as the male's bad cry is the typical multi-note song of the African Wood Owl (which in actuality is given by both sexes), and what they referred to as the female's cry is the single-note wail call (probably most often given by females, but possibly also by males and immature). They use the Lingala term esukulu ("es-oo-KOO-loo") to refer to "owl" but this really is a direct reference to the "bad" African Wood Owl. The term refers to both sexes of this species. They use the Lingala term lokio ("low-KEY-oh") to refer to the Vermiculated Fishing Owl (Scotopelia bouvieri), which they say (accurately) occurs only along the river (in riparian gallery forests), and that calls early morning, such as 3 to 5 a.m. Vermiculated Fishing Owls are not bad omens, as they never enter villages. They use the Lingala term enkimeli ("en-kim-EL-lee") to refer to Pel's Fishing Owl (Scotopelia peli), which also is a riparian or flooded-forest species, is not a bad omen, and does not enter villages.

They also spoke of another small owl that "has no cry", which might refer to a number of other possible small owls in the area that do not have calls like the wood owls or fishing owls (although they do have their own distinct, different vocalizations). In addition, they spoke of another owl-like creature called lyokokoli ("leeyo-ko-KOH-lee"; this is not Lingala but instead a local dialect, as there is no Lingala word for this) which is "very rare" and "appears once every forty to fifty years". It has a long wail call that increases in pitch. It is not an owl. They were unable to identify or describe it any further, and do not believe than any of the chiefs have personally seen this creature; it may exist in story only. The local belief in owls pertains to ndoki ("en-DOH-kee"), which is magic or sorcery.

When you hear an owl, you think of sorcery. Owls are feared because when one cries by your house it means bad news. Then you must throw stones at it to chase it away. Owls are bearers of news about the death of a relative elsewhere. They too knew the term esukulu.

In Serena Mountain Lodge on the slopes of Mount Kenya, the local people fear owls that "make noise", especially if the owl lands on the rooftop. This is particularly a bad omen, that someone will be reported having died the next day. The older generations have been told to believe such stories, but the younger generations, do not believe in such story telling.

At Lake Baringo in the Rift Valley of western Kenya, the Tugen tribe of the Kalenjin people, and the Mjemps tribe of the Masai people believes that if an owl appears on your house, it may signal that someone will die. If an owl appears on your house in the daytime, you may go blind. Such beliefs are held by the older generations but are now being replaced by more rational and conservation-minded thinking of the younger generation.

In the South Rim of the Grand Canyon in Arizona, pueblo tribes and the Hopi tribes in Northern Arizona and the Isleta which is south of Albuquerque along the Rio Grande River, New Mexico, owls are ere viewed as harbingers of ill health and ill fortune. If an owl comes to their house, shortly thereafter the younger children fall ill. If an owl was heard calling, siblings go outside and shout at it, try to compel it to leave.

Among Shuswap Tribe, it is believed that owls are forecasting messengers and usually portend a forthcoming death, but the messages are not always bad. You have to know how to read the messages, how to understand the owl's calls. It is only the larger owls that convey the messages of death, including Screech Owl, Great Horned Owl, and the others as well. The Chagga Tribe, east side of Mount Kilimanjaro, Tanzania, believes that:

If an owl comes to your window, it is very bad.

Women will pour sour milk onto the ground as a libation to ward off the owl.

Men will sacrifice live animals in sacred places, to ward off the owl.

If the owl returns, especially three times, people believe that it might be their ancestor. Then they will pour sour milk onto the ground and talk to the ancestors (kind of an appeasement libation), so that the ancestors/owl will not harm them.

If an owl lands on the roof of your hut or house, this is very bad and means death or illness.

Owls are the birds, which bring bad news mostly.

When the owl comes and sits on the roof, it brings bad news; it is the sign of a bad thing, especially if it also calls. If it sits there for a short time and flies away, it is OK; but not if it lingers.

If the owl perches on the roof and stays, and calls, how can this problem be solved? They mix red ochre and charcoal, and put it around the house and say, you have brought bad news here, just go away with the bad news, don't leave the bad news here.

If the owl perches in a nearby tree and calls, that can affect the cows. The cows may start dying. So toward this off, they put the red ochre and charcoal around the tree and tell the owl to go away with the bad news it has brought here.

The big owls are the ones that are easy to see, but all owls, including the small ones, bring bad news.

However, such beliefs are now fading, and occur only in rural pockets.

Owls in Mythology and Culture

Throughout history and across many cultures, people have regarded owls with fascination and awe. Few other creatures have so many different and contradictory beliefs about them. Owls have been feared and venerated, despised and admired, considered wise and foolish, and associated with witchcraft and medicine, the weather, birth and death. Speculation about Owls began in earliest folklore, too long ago to date, but passed down by word of mouth over generations.

In early Indian folklore, owls represent wisdom and helpfulness, and have powers of prophecy. This theme recurs in Aesop's fables and in Greek myths and beliefs. By the middle Ages in Europe, the owl had become the associate of witches and the inhabitant of dark, lonely, and profane places, a foolish but feared ghost. An owl's appearance at night, when people are helpless and blind, linked them with the unknown, its eerie call filled people with foreboding and apprehension: a death was imminent or some evil was at hand. During the 18th century, the zoological aspects of owls were detailed through close observation, reducing the mystery surrounding these birds. With superstitions dying out in the 20th century, in the West at least, the owl has returned to its position as a symbol of wisdom.

Owls in Greek and Roman Mythology

In the mythology of ancient Greece, Athene, the Goddess of Wisdom, was so impressed by the great eyes and solemn appearance of the owl that, having banished the mischievous crow, she honored the night bird by making him her favorite among feathered creatures. Athene's bird was a Little Owl, (Athene noctua). This owl was protected and inhabited in great numbers in the Acropolis. It was believed that a magical "inner light" gave Owls night vision. As the symbol of Athene, the owl was a protector, accompanying Greek armies to war, and providing ornamental inspiration for their daily lives. If an owl flew over Greek Soldiers before a battle, they took it as a sign of victory. The Little Owl also kept a watchful eye on Athenian trade and commerce from the reverse side of their coins.

In early Rome, a dead owl nailed to the door of a house averted all evil that it supposedly had earlier caused. To hear the hoot of an owl presaged imminent death. The deaths of Julius Caesar, Augustus, Commodus Aurelius, and Agrippa were apparently all predicted by an owl. "Yesterday, the bird of night did sit even at noonday, upon the market place, hooting and shrieking" (from Shakespeare's "Julius Caesar"). The Roman Army was warned of impending disaster by an owl before its defeat at

Charrhea, on the plains between the Euphrates and Tigris rivers. According to Artemidorus, a second Century soothsayer, to dream of an owl meant that a traveler would be shipwrecked or robbed. Another Roman superstition was that witches transformed into owls, and sucked the blood of babies.

In Roman mythology, Proserpine (Greek: Persephone) was transported to the underworld against her will by Pluto (Greek: Hades), god of the underworld, and was to be allowed to return to her mother Ceres (Greek: Demeter), goddess of agriculture, providing she ate nothing while in the underworld. Ascalpus, however, saw her picking a pomegranate, and told what he had seen. He was turned into an owl for his trouble - "a sluggish Screech Owl, a loathsome bird".

Owls in English Folklore

Folklore surrounding the Barn Owl is better recorded than for most other owls. In English literature, the Barn Owl had a sinister reputation probably because it was a bird of darkness, and darkness was always associated with death. During the 18th and 19th centuries, the poets Robert Blair and William Wordsworth used the Barn Owl as their favorite "bird of doom." During that same period, many people believed that the screech or call of an owl flying past the window of a sick person meant imminent death. The Barn Owl has also been used to predict the weather by people in England. A screeching owl meant cold weather or a storm was coming. If heard during foul weather a change in the weather was at hand. The custom of nailing an owl to a barn door to ward off evil and lightning persisted into the 19th century.

Another traditional English belief was that if you walked around an pwl in a tree, it would turn and turn its head to watch you until it wrung its own neck. Among early English folk cures, alcoholism was treated with owl egg. The imbiber was prescribed raw eggs and a child given this treatment was thought to gain lifetime protection against drunkenness. Owls' eggs, cooked until they turned into ashes, were also used as a potion to improve eyesight. Owl broth was given to children suffering from Whooping cough. Odo of Cheriton, a Kentish preacher in the 12th Century has this explanation of why the owl is nocturnal: The owl had stolen the rose, which was a prize awarded for beauty, and the other birds punished it by allowing it to come out only at night. In parts of northern England, it is good luck to see an owl.

Owls in American Indian Culture

Among the different American Indian tribes, there are many diverse beliefs regarding the owl. According to an Indian legend, the 'Spedis Owl' carving was placed on a rock to serve as a protector from the 'water devils' and monsters that could pull a person into the water. The owl on a rock may have also indicated the ownership of that location for fishing. This petroglyph, the 'Spedis Owl' was salvaged from along the Columbia River just before The Dalles Dam flooded the area in 1956. The carving is on display at Horsethief Lake State Park, Washington.

To an Apache Indian, dreaming of an owl at night signified approaching the death. Cherokee shamans valued Eastern Screech-Owls as consultants as the owls could bring on sickness as punishment. The Cree people believed Boreal Owl whistles were summons from the spirits. If a person answered with a similar whistle and did not hear a response, then he would soon die. The Dakota Hidatsa Indians saw the Burrowing Owl as a protective spirit for brave warriors. The Hopis Indians see the Burrowing Owl as their god of the dead, the guardian of fires and tender of all underground things, including seed germination. Their name for the Burrowing Owl is Ko'ko, which means "Watcher of the dark" They also believed that the Great Horned Owl helped their Peaches grow.

The Inuit believed that the Short-eared Owl was once a young girl who was magically transformed into an owl with a long beak. However, the owl became frightened and flew into the side of a house, flattening its face and beak. They also named the Boreal Owl "the blind one", because of its tameness during daylight. Inuit children make pets of Boreal Owls. The people of Northwest coast (Kwagulth) believed that owls represented both a deceased person and their newly released soul. The Kwakiutl Indians were convinced that Owls were the souls of people and should therefore not be harmed, for when an owl was killed the person to whom the soul belonged would also die. The Lenape Indians believed that if they dreamt of an owl it would become their guardian. The Menominee people believed that day and night were created after a talking contest between a Saw-whet Owl (Totoba) and a rabbit (Wabus). The rabbit won and selected daylight, but allowed nighttime as a benefit to the vanquished owl.

The Montagnais people of Quebec believed that the Saw-whet Owl was once the largest owl in the world and was very proud of its voice. After the owl attempted to imitate the roar of a waterfall, the Great Spirit humiliated the Saw-whet Owl by turning it into a tiny owl with a song that sounds like dripping water. To the Mojave Indians of Arizona, one would become an owl after death, this being and interim stage before becoming a water beetle, and ultimately pure air. According to Navajo legend, the creator, Nayenezgani, told the Owl after creating it - "...in days to come, men will listen to your voice to know what will be their future". California Newuks believed that after death, the brave and virtuous became Great Horned Owls.

The wicked, however, were doomed to become Barn Owls.

In the Sierras, native peoples believed the Great Horned Owl captured the souls of the dead and carried them to the underworld. The Tlingit Indian warriors had great faith in the owl; they would rush into battle hooting like Owls to give themselves confidence, and to strike fear into their enemies. A Zuni legend tells of how the Burrowing Owl got its speckled plumage: the owls spilled white foam on themselves during a ceremonial dance because they were laughing at a coyote that was trying to join the dance. Zuni mothers place an Owl feather next to a baby to help it sleep.

Owl Mythology A Global Perspective

According to an ancient Arabic treatise, from each female owl supposedly came two eggs, one held the power to cause hair fall out and one held the power to restore it. Arabs used to believe that the spirit of a murdered man continues to wail and weep until his death is avenged. They believed that a bird that they called "al Sada" (or the death-owl) would continue to hoot over the grave of a slain man whose death had not been avenged. The bird would continue to hoot endlessly until the slain man's death was avenged.

Arctic Circle: A little girl was turned into a bird with a long beak by magic, but was so frightened she flapped about madly and flew into a wall, flattening her face and beak. Therefore, the owl was created.

Australia: Aborigines believe bats represent the souls of men and owls the souls of women. Owls are therefore sacred, because your sister is an owl - and the owl is your sister.

Aztecs: One of their evil gods wore a Screech Owl on his head.

Babylon: Owl amulets protected women during childbirth.

Belgium: A priest offered the owl his church tower to live in, if the bird would get rid of the rats and mice that plagued his church.

Bordeaux: Throw salt in the fire to avoid the owl's curse.

Borneo: The Supreme Being turned his wife into an owl after she told secrets to mortals.

Brittany: Owl seen on the way to the harvest is the sign of a good yield.

Burma: During a quarrel among the birds, the owl was jumped upon and so his face was flattened.

Cameroon: Too evil to name, the owl is known only as "the bird that makes you afraid".

Carthage: The city was captured by Agathocles of Syracuse (Southern Italy) in 310 BC. Afterward, he released owls over his troops and they settled on their shields and helmets, signifying victory in battle.

Celtic: Owl was a sign of the underworld.

China: Owl is associated with lightning (because it brightens the night) and with the drum (because it breaks the silence). Placing owl effigies in each corner of the home protect it against lightning. The owl is a symbol of too much Yang (positive, masculine, bright, active energy).

Croatia: Owl is a symbol of City of Krk on the island of Krk, and is the protector of the island of Solta, where it is called "cuvitar".

Ethiopia: A man condemned to death was taken to a table on which an owl was painted, and then expected to take his own life.

Etruria: In the Etruscans of Ancient Italy, the owl was an attribute of the god of darkness.

France: When a pregnant woman hears an owl, it is an omen that her child will be a girl.

Germany: If an owl hoots as a child is born, the infant will have an unhappy life. "A charm against the terrible consequences of being bitten by a mad dog was to carry the heart and right foot of an owl under the left armpit." (Encyclopedia of Superstitions)

Greenland: The Inuit tribe sees the owl as a source of guidance and help.

Hawaii: Owls were featured in old war chants.

Incas: Owls are venerated for heir beautiful eyes and head.

India: Seizures in children could be treated with a broth made from owl eyes. Rheumatism pain was treated with a gel made from owl meat. Owl meat could also be eaten as a natural aphrodisiac. In northern India, if one ate the eyes of an owl, they would be able to see in the dark. In southern India, the cries of an owl were interpreted by its numbers. One hoot was an omen of impending death; two meant success in anything that would be started soon after; three represented a woman being married into the family; four indicated a disturbance; five denoted coming travel; six meant guests were on the way; seven was a sign of mental distress; eight foretold sudden death; and nine symbolized good fortune. In some parts of the Indian Subcontinent, people believed that the owl was married to the bat.

The Barn owl is the "vahana" (transport/vehicle/mount) of the Hindu goddess of wisdom, Lakshmi. As such, the owl is held as a symbol of wisdom and learning. The Ragle owls, especially the Rock eagle owl [Bubo bengalensis] and the Brown fish owl [Bubo zeylonensis] are called "ullu" in Hindi and the word is also used as a synonym for "idiot" or "imbecile". The most chilling sound during the quiet and cold winter nights in the plains of Bengal is perhaps the call of the "kaal penchaa", the Brown Hawk Owl. The rhythmic "kuk - kuk - kuk" is believed to be a foreboding of impending death.

Indonesia: Around Manado, on the isle of Sulawesi, people consider owls very wise. They call them 'Burung Manguni'. Every time if someone wants to travel, he listens to the owls. The owls make two different sounds; the first means it is safe to go, and the second means it is better to stay at home. The Minahasa, people around Manado, take those warnings very seriously. They stay at home when Manguni says so.

Illinois: Kill an owl and revenge will be visited upon your family.

Iran: In Farsi, the Little Owl (Athene noctua) is called "Joghde-kochek". It is said that this bird brings bad luck. In Islam, it is forbidden (Haram) to eat after seeing an owl.

Ireland: Owl that enters the house must be killed at once, for if it flies away it will take the luck of the house with it.

Israel: In Hebrew lore, the owl represents blindness and desolation and is unclean.

Jamaica: To ward off the owl's bad luck, cry "Salt and pepper for your mammy".

Japan: Among the Ainu people, the Eagle Owl is revered as a messenger of the gods or a divine ancestor. They would drink a toast to the Eagle Owl before a hunting expedition. The Screech Owl warns against danger. Though they think, the Barn Owl and Horned Owl are demonic. They would nail wooden images of owls to their houses in times of famine or pestilence.

Latvia: When Christian soldiers entered his temple, the local pagan god flew away as an owl.

Lorraine: Spinsters go to the woods and call to the owl to help them find a husband.

Louisiana: Owls are old people and should be respected. Louisiana Cajuns (individuals who share the French-based culture originally brought to Louisiana by exiles from the French colony of Acadia in the 18th century) thought you should get up from bed and turn your left shoe upside down to avert disaster, if you hear an owl calling late at night.

Luxembourg: Owls spy treasures, steal them, and hoard them.

Madagascar: Owls join witches to dance on the graves of the dead.

Malawi: Owl carries messages for witches.

Malaya: Owls eat newborn babies.

Mayarts: Owls were the messengers of the rulers of Xibalba, the Place of Phantoms.

Mexico: Owl makes the cold North wind (the gentle South wind is made by the butterfly). The Little Owl was called "messenger of the lord of the land of the dead", and flew between the land of the living and the dead.

Middle East: Owl is linked with destruction, ruin, and death. They are believed to represents the souls of people who have died un-avenged. Seeing an owl on the way to battle foretells a bloody battle with many deaths and casualties. Seeing an owl at somebody's house predicts his or her death. Seeing an owl in your sleep is fine as long as you do not hear its voice. An owl's sound forecasts a bad day. A person who nags and complains a lot is compared to an owl. When someone is grumpy or delivering bad news, he is said to have a face like an owl.

Mongolia: The Burial people hang up owl skins to ward-off evil.

Mongolia (Inner): Owls enter the house by night to gather human fingernails.

Morocco: The cry of owls can kill infants. According to Moroccan custom, an owl's eye worn on a string around the neck was an effective talisman to avert the "evil eye".

New Mexico: The hooting of owls warns of the coming of witches.

New Zealand: For the Maoris, the owl is an unlucky bird.

Newfoundland: The hoot of the Horned Owl signals the approach of bad weather.

Nigeria: In legend, Elullo, a witch and a chief of the Okuni tribe, could become an owl. In certain parts of Nigeria, natives avoid naming the owl, referring to it at "the bird that makes your afraid".

Persia: Wizards use arrows tipped with a bewitched man's fingernails to kill owls.

Peru: Boiled owl is said to be a strong medicine.

Poland: Polish folklore links owls with death. Girls who die unmarried turn into doves; girls who are married when they die turn into Owls. An owl cry heard in or near a home usually meant impending death, sickness, or other misfortune. An old story tells how the owl does not come out at during the day because it is too beautiful, and would be mobbed by other, jealous birds.

Puerto Rico: The owl is called "Mucaro". Back in the 1800s, the people from the mountain coffee plantations used to blame the little mucaro for the loss of coffee grains. The belief was that the coffee was part of the owls' diet, and many owls were killed. There are old folklore songs on the subject, one goes like this: "Poor Mucaro, you are a gentleman, you just want to eat a rat, then the rat set up a trap, he eats the coffee grains,and people blame you."

Romania: The souls of repentant sinners flew to heaven in the guise of a Snowy Owl.

Russia: Hunters carry owl claws so that, if they are killed, their souls can use them to climb up to Heaven. Tartar shamen of Central Russia could assume owl shapes. Kalmucks holds the owl to be sacred because one once saved the life of Genghis Khan.

Samoa: The people are descended from an owl.

Saxony: Wend people say that the sight of an owl makes childbirth easier.

Scotland: It is a bad luck to see an owl in daylight.

Shetland Isles: A cow will give bloody milk, if scared by an owl.

Siberia: Owl is a helpful spirit.

Spain: Owl was once the sweetest of singers, until it saw Jesus crucified. Ever since, it has shunned daylight and only repeats the words 'cruz, cruz' ('cross, cross').

Sri Lanka: Owl is married to the bat.

Sumeria: The goddess of death, Lilith, was attended by owls.

Sweden: Owl is associated with witches.

Tangiers: Barn Owls are the clairvoyants of the devil.

Transylvania: Farmers used to scare away owls by walking round their fields naked.

Ural Mountains: Snowy Owls were made to stay behind while other birds migrate as a punishment for deception.

U.S.A: If you hear an owl-cry, you must return the call, or else take off an item of clothing and put it on again inside out.

Wales: Owl heard among houses means an unmarried girl has lost her virginity. If a woman is pregnant and she alone hears an owl hoot outside her house at night then her child will be blessed. In Welsh mythology, Blodeuedd, a woman made from flowers, is cursed by her husband's uncle, turning her into an owl. "You are never to show your face to the light of day, rather you shall fear other birds; they will be hostile to you, and it will be their nature to maul and molest you wherever they find you."

Human Impacts on Owls

Humans have long roamed the earth interacting with other species and the habitat in which they are found. These interactions or impacts can be considered harmful or beneficial. There is concern because some of the impacts are harmful. Many people interpret the word "impact" as having a negative connotation. Impact will imply both positive and negative aspects of the interactions between humans and owls. This is necessary to clarify because humans cause both harmful and beneficial impacts to owls. One group of species being impacted by humans is owls composed of the families Tytonidae and Strigidae. Owls are fairly secretive and many people do not realize the impacts which humans impose on them.

There are 19 species of owls in the United States. Many people think that all owls live in trees. Some species such as the Burrowing Owl (Athene cunicularia) live predominantly underground and the Short-Eared Owl (Asio flammeus) nests directly on the ground. Not all owls are large. The Great Grey Owl (Strix nebulosa) stands approximately 74 cm in height, while other owls such as the Elf Owl (Micrathene whitneyi) stand only 15 cm in height. With such diversity, a single human interaction can have varying outcomes on different species.

Impacts to owls can occur at two levels: individual and population. The individual scale involves impacts, which effect only select individuals in a given geographic area. Most of the time, these impacts do not have a drastic effect on an overall population dynamics of owls in a geographic range. The population scale impacts are of more concern to biologists because they usually affect most if not all owls threatening their survival.

Individual Impacts

Some impacts only affect individual owls. Individual impacts against owls include deliberate shooting, trapping and poisoning. Although impacts such as these have decreased due to education and legislation, it continues to be a problem. Individuals who feel that the birds are posing a threat to their livelihood are more likely to take actions to eliminate the birds. There also have been reports of killings to collect the feathers or for sport. There have been documented killings of Spotted Owls in areas where logging is a vital part of the economy. There have also been reports of shootings of Long-Eared, Burrowing, and Short-Eared Owls. Since most species of owls are nocturnal, they are not facing impacts like diurnal bird species. There has been debate over the effect that these killings have on owl populations. Some researchers have concluded that these killings may cause declines in local populations of owls while others say there are no long-term impacts to the population as a whole.

Power lines are both beneficial and detrimental to owls. The beneficial impact that owls gain from power lines and poles is their availability for perching and roosting. The areas around power lines that are cleared provide good habitat for hunting. The owl can sit on the power line or pole and scope out the surrounding area for prey. The main detriment that power lines cause is electrocution. Various owl species have been found electrocuted including Great Horned Owls (Bubo virginianus) and Short-Eared Owls (Asio flammeus). Owls probably encounter power lines more often than reported. Some individuals may be injured and crawl away from the area never to be found. Owl populations probably are not significantly reduced due to electrocution. With power and utility technology increasing, negative impacts from power lines may continue to decrease.

Of all the individual impacts that owls face, automobiles appear to be the most detrimental. Many people drive at night when owls generally hunt. Many owls use highways and nearby areas for hunting. Owls perch in trees and other structures to watch for their prey. When an owl sees their prey, they move in for the kill flying close to the ground. Since owls have frontally situated eyes, they tend to have a narrow visual field of approximately 110 degrees resulting in "tunnel vision". If an owl is flying perpendicularly over a road, it may not be able to see oncoming automobiles. In a study done in Hawaii from 1992-1994, 81 Barn Owls (Tyto alba), and five Short-Eared Owls (Asio flammeus) were evaluated to determine their cause of death. Of the 40 that died from trauma, 29 appeared to have collided with automobiles. A study on Eastern Screech Owl (Otus asio) revealed that vehicles, while hunting near roads frequently, kill Screech Owls. The number of deaths by automobiles may not be considered detrimental to a specific population of owls.

Another cause of random killings of owls is barbed-wire fences. The number of owls killed by barbed-wire fences probably is not significant, but it has been reported many times. Throughout the United States, Great Horned, Burrowing, Short-Eared, and Barn Owls have been found entangled in fences. On 16 July 1993, a dead Long-Eared Owl (Asio otus) was discovered hanging from a barbed-wire fence in Colorado. Raptor rehabilitator Heather Best of the Western North Carolina Nature Center stated that a majority of the Great Horned Owls that they receive are injured from being entangled in barbed-wire fencing.

The list of human caused impacts that effect owls on the individual scale also include dams, research, railroads, airplanes, recreation, and mass noise pollution such as sonic booms. There have even been reports of owls being caught in abandoned fishing line. These individual impacts are detrimental to owls, but impacts that affect owls on a population scale are of more concern to scientist. Population scale impacts cause greater impact to owls, so they generally are studied more and are ecologically more important.

Toxicological Impacts on Owls

Chemical contamination is considered to be one of the most detrimental threats to owl populations. Whereas most human impacts on owls are on a local scale, pesticides and other chemicals affect owls on a global scale. Chemicals are not confined to any geographic area because they can travel through the air, soil, and water. One reason that chemical contamination is so detrimental to owls is because they cannot recognize and avoid pesticides. With other human impacts, the owls are able to recognize the changes and move to a different location or adapt to the changes. Owls are not able to avoid chemical pollution because it is a global problem.

Most of the time chemical contamination does not directly kill owls. Instead, chemicals usually affect owls through bioaccumulation. Lower levels of the food chain do not consume high levels of pesticides. As you move up the various trophic levels, the accumulation of chemicals increases. The species at the bottom of the food chain take in low concentrations of the chemical. When the predator eats the prey, the toxins from the prey accumulate in the predator. This build up of chemicals moves up the food chain. Since owls are at the top of the food chain, they tend to accumulate large amounts of the chemicals from their prey. Retention of the chemicals in their bodies is what causes the problems with bioaccumulation. A problem with many of the pesticides, rodenticides and other chemicals is that many of them are resistant to breakdown in the environment. Chemicals, which do not break down, can cause problems years down the road after they have been put in the environment. DDT is a good example of a resistant chemical that has affected owls and other raptors.

Scientists have realized the importance of studying the impacts that chemicals have on various birds of prey. Various methods can be used in determining pollutant levels in owls, such as the evaluation of road kills, infertile and fertile eggs, eggshell fragments, and performing feather, blood, fat, liver and muscle biopsies. Chemicals also cause secondary poisoning in owls. Owls receive secondary poisoning through prey, which have ingested a chemical that may or may not break down in the prey or owls body into another chemical (such as DDT breaking down into DDE). Studies on the effects of Famphur, an organophosphate, on Barn Owls suggest that Famphur is not considered as dangerous as other chemicals, because biological processes inactivate it rapidly. Due to the possibility of Famphur being accumulated prior to its breakdown and killing potential owl prey, it was assumed that there was a possibility that prey, which have been in contact with Famphur, may cause secondary poisoning to birds of prey. One study suggests that owls and other raptors that eat prey poisoned by Famphur could experience secondary poisoning.

Another instance of secondary poisoning was a study of the impact of Volid on owls. In this study, Volid, a rodenticide, was used to control the vole population in an apple orchard. Radio transmitters were attached to 38 Eastern Screech Owls, five Barred Owls, three Red-Tailed Hawks, two Great Horned Owls, and two Long-Eared Owls. After treatment of 57% of the orchard, six Screech Owls were positively identified as being killed by Volid. One Long-Eared Owl was found to have died from secondary poisoning. Many species of owls prey on small rodents such as voles. From eating voles poisoned by Volid, the study concluded that there are significant hazards to Eastern Screech Owls and other raptors due to the bioaccumulation of Volid.

There are recorded instances of direct killings to certain species of owls from chemicals. In Canada, Carbofuran has been blamed in the reduction of Burrowing Owl populations by being applied directly to nest sites. Carbofuran is used as a pesticide to kill insects, mites, and nematodes. It can be used against pests in field, fruit, and vegetable and forest crops. This pesticide is known to be highly toxic to birds because of secondary poisoning, but also because of the impact birds face with direct contact. In the United States, granular Carbofuran was banned on 1st September, 1994. It is still legal for use as a spray, but is classified as a Restricted Use Pesticide (RUP). With its wide usage throughout North America, it poses a potential hazard to owls.

Not all pollutants are from agricultural practices or rodenticides. A study conducted in the late 1980's found Lead from mining in Great Horned Owls and Western Screech Owls in Northern Idaho. Although no deaths were recorded and the concentrations in these two species of owls were relatively low, lead in high concentrations can affect owls negatively or even cause death. Some toxins occurring naturally in the environment are lethal. Mercury is a naturally occurring element that is found in most environments. Humans release mercury into the environment in large concentrations through industry. Large amounts of mercury are dangerous to owls.

Some studies revealed the impacts of chemicals onto the owls. Carbonyl-based insecticides destroy non-target moths, which may be critical prey early in breeding season for Flammulated Owls. These moths are one of the preys of Flammulated Owls. Predation on moths, which have been in contact with specific insecticides, may cause secondary poisoning to Flammulated Owls and other species of owls. Great Grey Owls prey primarily on small rodents, many that are being poisoned with rodenticides. Barn owls have residues of pesticides and eggshell thinning due to pesticides used in agricultural practices. Burrowing Owls nest on the ground so they are directly affected by application of insecticides, pesticides and other chemicals applied on the ground.

Agricultural Impacts on Owls

The primary way that agriculture affects owls is by habitat alteration. Farming and ranching are a primary land use in the United States. Since agricultural practices often modify landscapes, this can have both beneficial and harmful impacts on owls. The most beneficial aspects of farming to owls are the planting of trees and the building of structures such as barns. One species of owl that benefits from barns and other buildings on agricultural lands is the Barn Owl. The Barn Owl inhabits countryside with open fields and hedges for hunting and uses old buildings for breeding sites. Throughout the early-mid 20th century, many barns and farmhouses were built. These provided the habitat for Barn Owls to thrive in.

Grazing has both beneficial and detrimental impacts. In some circumstances, grazing may have beneficial impacts on Burrowing Owls. The grazing of lands with tall grasses can create suitable habitat for burrowing mammals. Humans therefore provide nesting sites for Burrowing Owls. There are two primary ways that owls are impacted by grazing. The first is the alteration of nesting site availability. "Intensive grazing in meadow and marsh habitats may adversely affect the nesting habitat of ground nesting raptors such as the Short-Eared Owl." In North Dakota, Short-Eared Owls tend to avoid grazed habitat.

The second way owls are impacted by grazing is through prey abundance and vulnerability. Abundance of small bird, mammal, and reptile populations is reduced because of habitat alteration. Grazing can reduce the vegetation cover. Prey species that require low levels of cover are favored over those that need dense or tall cover. Lower ground cover makes prey more easily seen and captured by owls. In South Dakota, young Great Horned Owls hunt mainly in areas with little or no cover.

There are also indirect impacts that livestock-grazing pose. One is the development of water ponds. Stock ponds in Arizona have provided conditions suitable for mesquite forests, which provided habitat for Long-Eared Owls, which were previously not there. Prior to 1920, the Barn Owl was extremely rare in Ohio. The clearing of the countryside in the 1920's and 1930's greatly increased the owl population. Since then, the Barn Owl population has declined overall. This decline has been correlated with the availability of grasslands. As the grasslands have declined, so has the Barn Owl population. The population increase began in the 1920's and the major decline started in the 1960's.

In Idaho, a study was performed on the influence of agricultural development on raptor species. The study was done along the Snake River Plain in Southern Idaho. From 1986-87, studies were conducted via road surveys. Of the 11 species of raptors found, two were owl species. The Burrowing Owls were the only resident that flourished in the agriculturally developed areas. The Short-Eared Owl was also seen during the study. They were less abundant in developed areas. Owls may have also been secondarily affected by a decline in prey densities. Differences in prey populations and vegetation between developed vs. undeveloped rangelands may have a drastic beneficial or detrimental impact on owl densities. Although the Burrowing Owl flourished in the study area, all other raptor species appear to have been negatively affected by agriculture. Agricultural practices cause habitat alterations, but agriculture is not the only way that habitats are altered.

Impacts of Habitat Alteration on Owls

Human alteration and destruction of habitat causes significant impacts on owls. Loss of habitat is the primary factor in the decline of raptor populations throughout the world. In addition to agriculture, logging, urbanization, recreation, energy, and mineral development are altering owl habitats throughout the United States. In the United States, 98% of the tall-grass prairies have been plowed, half of the wetlands drained, 90-95% of old-growth forests cut, and overall forest cover reduced by 33%. Many species of owls have already begun to feel the tightening grip of altered habitats.

In America, the owl habitat is altered by urbanization and building of man-made structures. Urban areas have a population greater than 2,500. Approximately 75% of Americans live in urban areas and the remaining 25% live in rural areas. This can be beneficial or harmful to owl populations. Owls could benefit from urbanization because it moves people into a condensed area and keeps them from being spread out. The city removes habitat in a small area, keeping other areas from being directly affected. Urbanization can also be harmful. Altering habitat drastically can make it unsuitable for most owl species. Some species such as Screech Owls and Great

Horned Owls are more tolerant of urbanization, but most species are negatively impacted.

Some owl species tend to be habitat specialists. They have specific habitat requirements in which they need to survive. Pima County, Arizona is considering designating habitat for Northern Pygmy Owls (Glaucidium gnoma). In an article published in July 1999, there was concern by builders. The concern is that only 16% of the county is taxable. Designating 731,000 acres of habitat as critical for the Pygmy Owl could be harmful to the economy. If the land is designated, private landowners will need a federal permit to develop their land. In addition, taxes may have to be increased to compensate the loss of revenue. This in turn may result in fewer people being able to afford housing in the area. One person stated, "We have to ask, are we putting the owl before the people?" This is a controversial problem throughout the United States and will likely continue to be for some time to come.

In California's Silicon Valley, the Burrowing Owl is being affected by development. Land prices in the area are high because of the computer industry, so developers are the only people able to purchase and develop the land. This alters Burrowing Owl habitat. There is an effort throughout the region to try to preserve as much habitat as possible to save the decreasing population of Burrowing Owls. For example, Mission College has implemented a plan to preserve some owl sites. They installed artificial burrows and keep the grass cut low which the owl prefers. Although conserving the Burrowing Owl habitat has been difficult, the region is in the developmental stages of implementing a plan to conserve the Burrowing Owls and their habitat.

The most well known habitat alteration that affects owls is logging. The impact of logging on each species varies. The method of harvesting trees can determine that impact. The four main harvesting methods are selective cutting, shelterwood cutting, seed-tree cutting, and clear-cutting. Selective cutting is the removal of intermediate-aged or mature trees in an uneven-aged forest. This harvesting style can be both beneficial and harmful to owls. It may be beneficial to owls that do not prefer dense forests and those that can live in both old and intermediate-aged forests. This style of harvesting does not remove all the trees and leaves the forest in an uneven aged state with young, middle-aged, and mature trees. It may be detrimental to owls that prefer a specific age group of trees. Some owls such as Great Horned Owls are generalists. They are able to live in a variety of habitats. Others like the Spotted Owl are specialists and need old-growth forest to survive.

Shelterwood cutting is another form of logging. This style of harvesting involves removing the mature trees usually in two cuttings. Most of the mature trees are removed in the first cut leaving some smaller trees to shelter the seedlings. When the seedlings are well established, the remaining trees are cut. This style of logging can have a drastic impact on owls that prefer mature or old growth forest. The benefit of this logging style is that it does not remove all the older trees at once. They are removed over a period, which may be beneficial in allowing owls to adapt to the altered landscape more easily.

Seed-tree cutting involves the removal of nearly all the trees in one cutting. A few of the best trees are left for producing seedlings. This style of cutting may be beneficial to species that do not require a developed forest because the trees are removed. The new forest is an even-aged forest. Spotted Owls and Great Grey Owls would most likely be harmed by such logging practices.

Clear-cutting removes all trees in a given area in one cutting. There are variations of clear-cutting such as strip logging and whole-tree harvesting. The cutting may be in a patch, strip, or a whole stand. In the United States, nearly 66% of the annual timber harvest is attained by clear-cutting. The practice of clear-cutting has a negative impact on owls that depend on trees for nesting. Clear-cutting can also be beneficial. It creates foraging habitat. Many owls such as Great Grey Owls prefer to hunt along forest margins, which provide open areas for visual searching for prey. In addition, clear-cutting can create suitable habitat for owls that do not nest in trees such as the Burrowing Owl and the Short-Eared Owl.

Studies on the Spotted Owl, Great Grey Owl, and the Barred Owl are presented in the following section to illustrate the impact of human-altered habitat on owl populations. These species have been chosen because of the abundance of information on them.

Spotted Owl: The Northern Spotted Owl has become the symbol in America with the Jobs/Logging vs. Owl debate. In one of the studies, 98% of the sites with Spotted Owls were either old-growth forests or a mixture of mature and old growth forests. They also prefer to live in multi-layered canopies and uneven-aged forests. The Spotted Owls do not build their own nest. They utilize natural sites of broken treetops and cavities. The average home ranges of Spotted Owls range from 400 to 2776 hectares. The main habitat alteration that seems to affect Spotted Owls is the logging of old-growth forest in the western United States. The concern for the Spotted Owl began back in the 1970's when the logging vs. habitat debate in the Pacific Northwest was on the rise. Some studies have indicated that Spotted Owl populations have decreased with the elimination of old-growth forests and the replacement by younger forests. For many decades throughout the 20th century, the Pacific Northwest was heavily logged because of the ever-increasing need for timber in the United States. Much of the old-growth forests in the United States have vanished, but most of what remains is located in the Northwest. Less than 10% of the old-growth forests that once stood in the Northwest are still standing. Many people in the Northwest depend on logging for employment, so it is understandable that there is much opposition to the preservation of old-growth forests for Spotted Owls.

Although there have been some concerns about the effect that preservation of owl habitat may have on jobs and the economy, the economy in Oregon is currently on the rise. In 1993, President Clinton's forest plan caused many job losses due to lumber trade regulations, but it included occupational retraining. Timber harvesting is not the only reason for concern with the Spotted Owl. The California Spotted Owl is also losing habitat for breeding and foraging. New residential developments, fires, recreational activities, and the "mining" of water from streams in the owl's habitat are influencing the California Spotted Owls population. This species of owl will continue to be one of concern in the United States because of their dwindling numbers.

Great Grey Owl: The Great Grey Owl prefers mature or old growth forests throughout the Pacific Northwest. They nest in abandoned nests or on top of tree trunks. Logging effects on Great Grey populations can be either beneficial or harmful. Depending on the extent, type, and time of logging, it can either benefit or be detrimental to the species. In northeastern Oregon, owls spend the majority of their time foraging in selectively logged stands, and in southeastern Idaho, the owls prey on gophers in clear-cut areas. The reduction of dead trees used for nesting, and dense canopy trees used by juveniles are problems that logging causes with Great Grey Owls.

Barred Owl: The Barred Owl lives primarily in mature forests and swamps. They nest in tree cavities but sometimes will use abandoned nests. Logging of mature and old growth forests has been the primary habitat alteration that affects this species. In Connecticut and New Hampshire, Barred Owls avoid developed areas. However, there is evidence that Barred Owls are able to sustain human development more readily than other owls. Although there are many other species of owls, this is a sample of some of the impacts caused by habitat alteration and destruction.

Legal Protection to Owls

One of the beneficial impacts humans have on owls is through legal protection. In the United States, there are serious penalties for harming all birds of prey including owls. Legislation has helped to increase public awareness, and reduce shootings, poisoning, and habitat destruction. An early law, the Migratory Bird Treaty Act of 1918, prohibited the shooting of all raptors. Though it took many years for the public to become conscious of the act, it has had a positive impact on owls and other raptors throughout the United States. In 1973, the United States Congress approved the Endangered Species Act. This law restricts the possession of any endangered or threatened species in the United States. Not only does this include species found throughout the United States, but also those found worldwide. The ban on DDT was an extremely important piece of legislation. This chemical has had an impact on raptor species reproduction throughout the United States since its use began during WWII. Without this federal legislation, the populations of owls throughout the nation would be far more negatively impacted by deaths. Further legislation is needed to protect owls from habitat alteration, poisoning, and other negative human impacts.

Humans affect owls in many ways. Some impacts are beneficial while others are harmful. Some of the impacts only affect a few individual owls, while others may affect the entire population of a particular geographic region. Scientist to find a way to minimize or eliminate them is addressing harmful impacts. Overall, the status of owls in the United States is rather stable. Most owls are remaining steady in numbers while selective species like the Spotted Owl are declining in number. The outlook for owls in the future is a bright one. Now that the United States population is being exposed to environmental issues, the general population will most likely want to protect their natural resources and the species that live in them such as owls. Much of the study of raptor biology and human impacts on raptors is a fairly a new area of research. It is probable that impacts humans have on owls and other raptor species will continue to be of great concern. Humans are continuously altering owl habitat. With continued population growth and development, owls will continue to be at risk. Continued research is needed to help protect owls from the impacts of humans to maintain the stable population of owls.

Owls as Environmental Indicators

In China, owls have been associated with thunder and the summer solstice. Elsewhere, owls associated with old forests have been seen more recently by modern natural resource management agencies as prognosticators of the health and fate of such environments. These owls have been called management indicator species by agencies such as USDA Forest Service, who has identified the Northern Spotted Owl (Strix occidentalis caurina) as such an indicator of old-growth conifer forests of northwestern United States.

Nevertheless, the Spotted Owl has other cousins that have been, or can be, used to indicate the health of the vanishing ancient forests of the world. For example, in the greater Indian subcontinent ranges the Brown Wood Owl, Strix leptogrammica, which likewise inhabits mostly old and lesser disturbed forests of Sal (Shoria robusta). In the Himalayas, reside the Bay Owl (Phodilus badius), a rare denizen and indicator of dense evergreen submontane forests of cedar and other conifers. In the Russian Far East, the endangered Blakiston's Fish Owl (Ketupa blakistoni) also serves this function, found thinly scattered only in the increasingly rare dense riparian forests of old fir, larch, pine, and hardwoods in southern Siberian Ussuriland. Throughout the world, 82 owl species are closely associated with old forests, most awaiting recognition as useful indicators of old-forest conditions.

Towards a Tolerant Conservation

Owl mythologies have come virtually full circle in Europe and America. From the worst bird in the world, the owl has become almost the most popular. In addition, old mythologies actually make owl conservation easier. The old bad news has become a way of making owls appealing for a contemporary audience.

This transformation in the owl's image is yet to be fully researched, but we can offer some initial comparisons. Firstly, understanding European and American owl myths may help us better understand or interpret contemporary African and Asian attitudes. It also may help us understand owl taboos amongst today's Native Americans and Canadians, tribal South Americans, and other First Nations. Secondly, if owl mythologies have evolved so dramatically in the West, then perhaps they offer an insight into the way owl mythologies could eventually metamorphose in other parts of the world, such as in Africa and Asia. Thirdly, by understanding the patterns of owl mythology in modern day Africa and possibly South America and Asia, we might be able to understand our own cultural past in a better way.

Overall, the contemporary conservation community has not grasped the deeply negative image of owls in less developed parts of the world. In fact, conservationists in general tend to think of birds or other wildlife mostly or only in terms of their own Western ecological-science-based and conservation-oriented system. By failing to appreciate other patterns of belief about birds, they are actually putting themselves at a disadvantage. In the case of the owl, this is especially significant; because the overwhelming nature of owl beliefs is that the beast is evil. Even in Western cultures, until the 1950s owls were routinely nailed to barn doors in France and U.K. to ward off lightening and the evil eye. There is even evidence to suggest that these practices continue today in parts of rural Britain.

It is important not just to understand historical views of owls in myth and culture, but also how such views have changed over time to the present and travelled geographically as human cultures have moved. A major problem with bird folklore is that some texts repeat the same ideas without any critical analysis of the truth of what they say. Most of these ideas about owls probably refer to the early modern period and were almost certainly gathered in the late 19th to early 20th century. Now, many of them have completely died out, depending on the cultural development of the people in question. For example, "The Ainu and Their Folklore", analyzes this northern Japanese community's beliefs about animals, especially owls, at the end of the 19th century. These old cultural ideas sometimes are still retailed as if they were living traditions, when, in fact, modern Ainu may consider their ancestors owl beliefs as no more than old wives' tales, a part of their own colorful but quaint past. We should therefore attempt to weed out the living from the archaic. However, in Africa deep taboos about owls are still powerful and living traditions. In fact, in West Africa the pattern of owl beliefs probably reflects European attitudes up until the medieval age.

Miscellany

Rodent Control

Encouraging natural predators to control rodent population is a natural form of pest control, along with excluding food sources for rodents. Placing a nest box for owls on a property can help control rodent populations (one family of hungry barn owls can consume more than 3,000 rodents in a nesting season) while maintaining the naturally balanced food chain.

Attacks on Humans

Although humans and owls frequently live together in harmony, there have been incidents when owls have attacked humans. For example, in January 2013, a man from Inverness, Scotland suffered heavy bleeding and went into shock after being attacked by an owl, which was likely a 50-centimetre-tall (20 inches) eagle owl. The photographer Eric Hosking lost his left eye after attempting to photograph a tawny owl, which inspired the title of his 1970 autobiography 'An Eye for a Bird'.

Conservation Issues

All owls are listed in Appendix II of the international CITES treaty (the Convention on Illegal

Trade in Endangered Species of Wild Fauna and Flora). Although owls have long been hunted, a 2008 news story from Malaysia indicates that the magnitude of owl poaching may be on the rise. In November 2008, TRAFFIC reported the seizure of 900 plucked and "oven-ready" owls in Peninsular Malaysia. According to Chris Shepherd, Senior Program Officer for TRAFFIC's Southeast Asia office, "This is the first time we know of where 'ready-prepared' owls have been seized in Malaysia, and it may mark the start of a new trend in wild meat from the region. We will be monitoring developments closely." TRAFFIC commended the Department of Wildlife and National Parks in Malaysia for the raid that exposed the huge haul of owls. Included in the seizure were dead and plucked barn owls, spotted wood owls, crested serpent eagles, barred eagles, and brown wood owls, as well as 7,000 live lizards.

Conclusion

By incorporating the owls, once again into our modern cultures and by understanding the roles, they have played in diverse societies throughout the world and do play in ecosystems, we can admit and proffer their legitimacy as denizens of those environments we otherwise seek to exploit. Owls have served as marvelous and fantastic symbols of recreation, aesthetics, art, science, lore, political power, ethics, and even death. In the case of owls, the deep fears and anxieties they generated and the prophetic status they once held, and still hold in some cultures, present environmentalists with a handle with which to engage the interest and sympathies of a wider audience. By inviting owls into a full cultural circle, we can build a more tolerant understanding of all societies and ages, and incorporate wildlife conservation into the broader tapestry of human endeavors.

Bibliography

Ali, S. (1987) Indian Hill Birds. Oxford University Press, Bombay, India. 188 pp.

Ali, S., S.D. Ripley (1987) Compact Handbook of The Birds of India and Pakistan. Oxford University Press, Bombay, India. 816 pp.

Allen, G.T. (1990) A review of Bird Deaths on Barbed-Wire Fences. Wilson Bulletin. 102: 553-58.

Alvarenga, H.M.F., Höfling, E. (2003) Systematic revision of the Phorusrhacidae (Aves: Ralliformes). Papéis Avulsos de Zoologia. 43 (4): 55-91.

Anonymous (1987) Dictionary of Native American Art Symbols. Rock Art Research Education.

Arkinstall, R. (1994) Pesticide implicated in owl decline. Alternatives. 20 (3): 9.

Armstrong, E.A. (1958) The Folklore of Birds. Collins, London.

Austin, O.L. (1948) The Birds of Korea. Bulletin of the Museum of Comparative Zoology of Harvard College. 101:1-301.

Bachmann, T., Klän, S., Baughmgartner, W., Klaas, M., Schröder, W., Wagner H. (2007) Morphometric characterisation of wing feathers of the barn owl Tyto alba pratincola and the pigeon Columba livia. Frontiers in Zoology. 4: 23.

Batchelor, J. (1901) The Ainu and their Folklore. Religious Tract Society, London, U.K.

Beck, T.W., Gould Jr., Gordon, I. (1992) Background and the Current Management Situation for the California Spotted Owl. In: Verner, Jared, McKelvey, Kevin S., Noon, Barry R., Gutierrez, R.J., Gould, Gordon I., Beck, Thomas W. [Technical Coordinators]. The California Spotted Owl: A Technical Assessment of its Current Status. General Technical Report PSW-GTR-133. Albany, CA: Pacific Southwest Research Station, Forest Service, U.S. Department of Agriculture. Pp. 37-51, 60-1, 95-6.

Big Owl (Owl-Man) A Malevolent Apache Monster. Native-languages.org.

Blakiston's Fish Owl Project (2013) Fishowls.com

Bourke, J.G. (1958) An Apache Campaign in the Sierra Madre. Charles Scribner's Sons, New York, N.Y.

Browne, V. (1995) Animal Lore & Legend: Owl. Scholastic.

Campbell, W. (1994) Know Your Owls. Axia Wildlife.

Carbofuran. In Extension Toxicology Network Pesticide Information

Profiles.http://ace.orst.edu/info/extoxnet/pips/carbofur. htm Revised 1996.

Cenzato, E., Samtopitro, F. (1990) (English Translation Copy 1991). Owls: Art Legend History. Bullfinch Press, Littlebrown & Co., London and Canada.

Chopra, P. (2017) Vishnu's Mount: Birds in Indian Mythology and Folklore. Notion Press. p. 109. ISBN 978-1-948352-69-7

Cline, W. (1938) Religion and World View. In: The Sinkaietk or Southern Okanagon of Washington. General Series In Anthropology 6, Menasha, Wisconsin.

Cocker, M. (2000) African birds in traditional magico-medicinal use - A preliminary survey. Bulletin of the African Bird Club. 7 (1): 60-66.

Colvin, B.A. (1985) Common Barn-Owl Population Decline in Ohio and the Relationship to Agricultural Trends. Journal of Field Ornithology. 56 (3): 224-235.

Cuando el tecolote canta, el indio muere (2008) La Cronica

Deacy, S., Villing, A. (2001) Athena in the Classical World. Koninklijke Brill NV, Leiden. The Netherlands. ISBN 9004121420

Devine, A., Smith, D.G. (1985) Eastern Screech Owl (Otus asio) Mortality in Southern Connecticut. Connecticut Warbler. 5: 47-48.

Dubow, C. (2004) Hangover Cures. Forbes.

Dyson, M.L., Klump, G.M., Gauger, B. (1998) Absolute hearing thresholds and critical masking ratios in the European barn owl: A comparison with other owls. Journal of Comparative Physiology. 182 (5): 695-702.

Enriquez, P.L., Mikkola, H. (1997) Comparative study of general public owl knowledge in Costa Rica, Central America and Malawi, Africa. Pp. 160-166 In: J.R. Duncan, D.H. Johnson, T.H. Nicholls, Eds. Biology and Conservation of Owls of the Northern Hemisphere. General Technical Report NC-190, USDA Forest Service, St. Paul, Minnesota. 635 pp.

Eurasian Eagle Owl – Bubo bubo – Information, Pictures, Sounds. Owlpages.com

Eurasian Eagle Owl (2013) Oiseaux-birds.com.

Fienup-Riordan, A. (1996) The Living Tradition of Yup'ik Masks. University of Washington Press, Seattle. 320 pp.

Fitzner, R.E. (1975) Owl Mortality on Fences and Utility Lines. Raptor Research. 9: 55-57.

Fleay, D. 1968. Night Watchman of the Bush and Plain: Australian Owls and Owl-like Birds. Jacaranda Press.

Flood, J. (1997) Rock Art of the Dreamtime. Harper Collins Publishers, Sydney, Australia. 372 pp.

Forsman, E.D., Meslow, E. Charles, W., Howard, M. (1984) Distribution and Biology of the Spotted Owl in Oregon. The Journal of Wildlife Management. 48 (Suppl. 87): 1-64.

Galeotti, P., Diego, R. (2007) Head ornaments in owls: what are their functions? Journal of Avian Biology. 38 (6): 731-736.

Gimbutas, M. (2001) The Living Goddesses. University of California Press, p. 158. ISBN 0520927095

Glick, D. (1995) Having Owls and Jobs Too. National Wildlife. National Wildlife Federation. Pp. 8-13.

Glory of the Morning, Hocąk Encyclopedia.

Gore, M.E.J., Won, P. (1971) The Birds of Korea. Royal Asiatic Society, Korea Branch, Seoul, Korea.

Gup, T. (1990) Owl vs. Man. Time Magazine. Pp. 56+. 25 June 1990.

Gutierrez, R.J., Franklin, A.B., Lahaye, W.S. (1995) Spotted Owl (Strix occidentalis). In: Poole, A., Gill, F. [Eds.]. The Birds of North America. No.179. Philadelphia : The Academy of Natural Sciences , Washington, D.C. : The American Ornithologists' Union.

Haaramo, M. (2006) Mikko's Phylogeny Archive: Caprimulgiformes – Nightjars. Mortimer, Michael (2004) The Theropod Database: Phylogeny of taxa.

Harrison, C., Greensmith, A. (1995) Koko maailman linnut (in Finnish). Translated by Laine, Lasse J., Nikander, Pekka. Helsinki Media. p. 198. (ISBN: 951-875-637-6)

Haug, E.A., Millsap, B.A., Martell, M.S. (1993) Burrowing Owl (Speotyto cunicularia). In: Poole, A., Gill, F. [Eds.]. The Birds of North America. No.61. Philadelphia :

The Academy of Natural Sciences , Washington, D.C. : The American Ornithologists' Union.

Hegdal, P.L, Colvin, B.A. (1988) Potential Hazard to Eastern Screech-Owls and Other Raptors of Brodifacoum Bait Used For Vole Control in Orchards. Environmental Toxicology and Chemistry. 7: 245-260.

Henny, C.J., Blus, L.J., Hoffman, D.J., Grove, R.A. (1994) Lead in Hawks, Falcons and Owls Downstream from a Mining Site on the Coeur d'Alene River, Idaho. Environmental Monitoring and Assessment. 29: 267-288.

Hill, E.F, Mendenhall, V.M. (1980) Secondary Poisoning of Barn Owls with Famphur, an Organophosphate Insecticide. Journal of Wildlife Management. 44: 676-81.

Hiren, S., J. Joshua and J. Pankaj (2003) Nest of Ashy Crown Finch Lark (Eremopteryx grisea Ticeh.) (Scopoli) in Bhachau Taluka, Kachchh District, Gujarat. Newsletter for Birdwatchers. 43 (1): 13. (ISSN: 0028-9426)

Holmes, B.. (1998) City Planning for Owls. National Wildlife. 36 (6): 46.

Holmgren, V.C. (1988) Owls in Folklore and Natural History. Capra Press, Santa Barbara, CA. 175 pp.

Hughes, A. (1979) A schematic eye for the rat. Vision Res. 19 (5): 569-588.

Ian, H. (2007) Ask an expert: Are barn owl feathers waterproof? RSPB.

Ingersoll, E. (1923) (Reprinted 1958) Birds in Legend, Fable and Folklore. Longmans, Green and Co., London.

International Science & Engineering Visualization Challenge: Posters & Graphics 2013) Science. 339 (6119): 514-515. doi:10.1126/science.339.6119.514.

Johnsgard, P.A. (1988) North American Owls: Biology and Natural History. Smithsonian Institute Press, Washington.

Joshi, P.N. and H.B. Soni (2019) Biodiversity Conservation by Local Communities of Kutch, Gujarat, India. India Wilds Newsletter. 11 (3): 8-13. (ISSN: 2394-6946)

Joshua, J. and H.B. Soni (2005) Occurrence of Franklin's or Allied or Savanna Nightjar (Caprimulgus affinis) in Bhuj Taluka, Kachchh District, Gujarat. Newsletter for Birdwatchers. 45 (4): 61. (ISSN: 0028-9426)

Joshua, J., H. Soni, N.M. Joshi, P.N. Joshi and O. Deiva (2005) Occurrence of Grey-headed Canary Flycatcher (Culicicapa ceylonensis) in Jamnagar District, Gujarat, India. Journal of Bombay Natural History Society. 102 (3): 340-341. (ISSN: 0006-6982)

Joshua, J., H. Soni, N.M. Joshi, P.N. Joshi and S.V.S. Rao (2005) Sighting of Common or Collared Sand Martin (Riparia riparia) and Plain or Indian Sand Martin (Riparia paludicola) in Banaskantha District, North Gujarat, India. Journal of Bombay Natural History Society. 102 (2): 233. (ISSN: 0006-6982)

Joshua, J., H.B. Soni, N.M. Joshi and P.N. Joshi (2005) Sighting of Fieldfare (Turdus pilaris) in Firozpura of

Banaskantha District, North Gujarat, India. Newsletter for Birdwatchers. 45 (4): 59. (ISSN: 0028-9426)

Joshua, J., S.F.W. Sunderraj, V. Gokula and H.B. Soni (2008) Status and Conservation of Some Threatened Faunal Biodiversity of Arid Kachchh. In: Forest Biodiversity (Vol. I) (Eds. K. Muthuchelian, S. Kannaiyan and A. Gopalam), Associated Publishing Company, New Delhi. 272 pp. (ISBN: 81-85211-76-0)

Juvonen, A., Muukkonen, T., Peltomäki, J., Varesvuo, M. (2009) Linnut vauhdissa (in Finnish). Tammi. p. 178, 187. ISBN 978-951-31-4604-7

Keats, J. (1884) Hyperion in 'The Poetical Works of John Keats'. Bartleby.com

Keyser, J.D., Pedersen, C., Bettis, G.M., Poetschat, G., Hiczun, H. (1998) OWL CAVE. Oregon Archaeological Society Publication No. 11. Portland, Oregon. 116 pp.

Knowling, P. (1998) A Wisdom of Owls. Avenue Press.

Kochert, M.N, Millsap, B.A, Steenhof, K. (1988) Effects of Livestock Grazing on Raptors with Emphasis on the Southwestern U.S. 1988. In: Glinski, R.L, Pendleton, B.G, Moss, M.B, Lefranc Jr, M.N., Millsap, B.A., Hoffman, S.W. [Eds]. Proceedings of the Southwest Raptor Management Symposium and Workshop. 21-24 May 1986, University of Arizona, Tucson. National Wildlife Federation, Washington, DC, National Wildlife Federation Scientific and Technical Series No. 11. pp. 325-334.

König, C., Friedhelm, W., Jan-Hendrik, B. (1999) Owls: A Guide to the Owls of The World. Yale Univ Press. ISBN 0300079206

Konig, C., Welck, F., Jan-Hendrik, B. (1999) Owls: A Guide to the Owls of the World. Yale University Press, ISBN 978-0-300-07920-3.

Krüger, O. (2005) The evolution of reversed sexual size dimorphism in hawks, falcons and owls: A comparative study. Evolutionary Ecology. 19 (5): 467-486.

Kumar, J.I.N., H. Soni and R.N. Kumar (2007) Patterns of Site-Specific Variation of Waterbirds Community, Abundance and Diversity in relation to Seasons in Nal Lake Bird Sanctuary, Gujarat, India. International Journal of Bird Populations. 8: 1-20. (ISSN: 1074-1755) (USA)

Kumar, J.I.N., R.N. Kumar and H.B. Soni (2020) Ecological and Environmental Science: A Research Perspective (Google Play Books Edition). Google Book Publisher (GBP), USA. 825 pp. (GGKEY: KX6WDN8ZDBP)
https://play.google.com/store/books/details?id=UCf-DwAAQBAJ

Kumar, J.I.N., R.N. Kumar, B.S. Hiren Kumar and A.N. Gohil (1998) A Survey of Avifauna of Khatana and Waghai Forests of Gujarat. Pavo. 36 (1 & 2): 69-75. (ISSN: 0031-3297)

Larco, H., Rafael, H., Berrin, K. (1997) The Spirit of Ancient Peru Thames and Hudson. New York, ISBN 0500018022

Lenders, E.W. (1914) The Myth of the 'Wah-ru-hap-ah-rah,' or the Sacred Warclub Bundle. Zeitschrift für Ethnologie. 46: 404-420.

Leptich, D.J. (1994) Agricultural Development and its Influence on Raptors in Southern Idaho. Northwest Science. 68: 167-171.

Lipton, J. (1991) An Exaltation of Larks. Viking. ISBN 978-0-670-30044-0.

Lockwood, W.B. (1993) The Oxford Dictionary of British Bird Names. Oxford University Press, Oxford.

Long, K. (1998) Owls: A Wildlife Handbook. Johnson Books.

Lundberg, A. (1986) Adaptive advantages of reversed sexual size dimorphism in European owls. Ornis Scandinavica. 17 (2): 133-140.

Malek, S.S., S.A. Saiyed and H.B. Soni (2020) Population Dynamics and Habitat Preference of Sarus Crane (Grus antigone L.) in Selected Pockets of Central Gujarat. International Journal of Creative Research Thoughts. 8 (7): 1769-1775. (ISSN: 2320-2882)

Man needed hospital treatment after owl attack. Daily Telegraph (London) (25 January 2013)

Marcot, B.G. (1993) Conservation of Forests of India: An Ecologist's Tour. USDA Forest Service, Pacific Northwest Research Station, Portland OR. 127 pp.

Marcot, B.G. (1995) Owls of Old Forests of the World. USDA Forest Service General Technical Report PNW-GTR-343. Portland, Oregon, USA. 64 pp.

Marcot, B.G., Cocker, P.M., Johnson, D.H. (2000) Owls in Lore and Culture. Presented at Owls: The Biology, Conservation, and Cultural Significance of Owls. International Conference. Canberra, Australia. , 19-23 January 2000.

Marcot, B.G., Johnson, D.H. (2003) Owls in Mythology and Culture. Pp. 88-105. In: J. R. Duncan. Owls of the World: Their Lives, Behavior and Survival. Key Porter Books, Ltd., Toronto, Canada. 319 pp.

Marti, C.D. (1974) Feeding Ecology of Four Sympatric Owls . The Condor. 76 (1): 45-61. Einoder, Luke D. Alastair, M.M. Richardson (2007) Aspects of the Hindlimb Morphology of Some Australian Birds of Prey: A Comparative and Quantitative Study. The Auk. 124 (3): 773–788.

Marti, C.D. (1992) Barn Owl. In: Poole, A., Stettenheim, P., Gill, F. [Eds.]. The Birds of North America, No. 1. Philidelphia: The Academy of Natural Sciences, Washington, D.C: The Academy Ornithologists' Union.

Martin, D.J. (1973) Selected Aspects of Burrowing Owl Ecology and Behavior. The Condor. 75 (4): 446–456.

Martin, G.R. (1982) An owl's eye: Schematic optics and visual performance in Strix aluco L. J. Comp Physiol. 145 (3): 341-349.

Martin, L.C. (1996) Folklore of Birds. Globe Pequot Printers.

Mayr, G. (2005) Old World phorusrhacids (Aves, Phorusrhacidae): A new look at Strigogyps (Aenigmavis) sapea (Peters 1987). PaleoBios (Berkeley). 25 (1): 11–16.

McCallum, D.A. (1994) Flammulated Owl (Otus flammeolus). In: Poole, A., Gill, F. [Eds.]. The Birds of North America. No. 93. Philadelphia : The Academy of Natural Sciences, Washington, D.C. : The American Ornithologists' Union.

McGaa, E. (1990) Mother Earth Spirituality - Native American Paths to Healing Ourselves and Our World. Harper Collins Publishers, New York. 230 pp.

Medlin, F. (1967) (3rd Reprint, 1971) Centuries of Owls. Silvermine Publishers, Inc. 93 pp.

Mictlantecuhtli. Ancient History Encyclopedia.

Mikkola, H. (1997a) General public owl knowledge in Malawi. Soc. Malawi J. 50 (1): 13-35.

Mikkola, H. (1997b) Comparative study on general public owl knowledge in Malawi and in eastern and southern Africa. Nyala. 20: 25-35.

Miller, G.T. (1994) Living in the Environment: Principles, Connections, and Solutions. 8th Ed. Wadsworth Publishing Company, CA.

Mueller, H.C. (1986) The evolution of reversed sexual dimorphism in owls: An empirical analysis of possible selective factors . The Wilson Bulletin. 19 (5): 467.

Native American Indian Owl Legends, Meaning and Symbolism from the Myths of Many Tribes (2008) Native-languages.org.

Nengminza, D.S. (1996) The School Dictionary Garo to English. 13th Edition. Garo Hills Book Emporium, Tura, Meghalaya, India. 268 pp.

Neuhaus, W., Bretting, H., Schweizer, B. (1973) Morphologische und funktionelle Untersuchungen über den, lautlosen Flug der Eulen (Strix aluco) im Vergleich zum Flug der Enten (Anas platyrhynchos). Biologisches Zentralblatt. 92: 495–512.

Norberg, R.A. (1977) Occurrence and independent evolution of bilateral ear asymmetry in owls and implications on owl taxonomy. Philosophical Transactions of the Royal Society of London. Series B, Biological Sciences. 280 (973): 375-408.

Olson, S.L. (1985) The fossil record of birds. In: Farner, D.S., King, J.R. & Parkes, Kenneth C. (Eds.): Avian Biology 8: 79-238 (131, 267). Academic Press, New York.

Opler, M. (1965) An Apache Life-way. Cooper Square, New York.

Owl mystery unraveled: Scientists explain how bird can rotate its head without cutting off blood supply to brain (2013) Johns Hopkins Medicine.

Owl Pellets in the Classroom: Safety Guidelines. carolina.com

Owls in Lore and Culture. The Owl Pages. p. 3.

Owls: Hocąk Encyclopedia.

Peakall, D.B. (1987) Impacts and Mitigation Techniques. In: Pendleton, B.A, Millsap, B.A, Cline, K.W, Bird, D.M. [Ed]. Raptor Management Techniques Manual. National Wildlife Federation, Washington, DC., National Wildlife Federation Scientific and Technical Series No. 10. Pp. 321-329.

Peters, D.S. (2007) The fossil family Ameghinornithidae (Mourer-Chauviré 1981): A short synopsis. Journal of Ornithology. 148 (1): 25-28.

Postovit, H.R, Postovit, B.C. (1987) Impacts and Mitigation Techniques. In: Pendleton, B.A, Millsap, B.A, Cline, K.W, Bird, D.M. [Ed]. Raptor Management Techniques Manual. National Wildlife Federation, Washington, DC., National Wildlife Federation Scientific and Technical Series No. 10. Pp. 183-213.

Rackham, H. (1997) Pliny Natural History: Books 8-11. Harvard University Press, Cambridge, Mass.

Radin, P. (1990 [1923]) The Winnebago Tribe. Lincoln: University of Nebraska Press, pp. 7-9. ISBN 0803257104

Ray, V.F. (1939) Cultural Relations in the Plateau of Northwestern America. Fredrick Webb Hodge Anniversary Publication Fund 3.

Saiyed, S.A., S.S. Malek and H.B. Soni (2020) Local People Perception towards Conservation and Management of Sarus Crane (Grus antigone antigone L.) in Central Gujarat Region. International Journal of Current Microbiology and Applied Sciences. 9 (8): 2592-2604. (ISSN: 2319-7692)

Sánchez, M.A. (2004) Avian zoogeographical patterns during the Quaternary in the Mediterranean region and paleoclimatic interpretation. Ardeola. 51 (1): 91–132.

Saunders, N.J. (1995) Animal Spirits. MacMillan, London. 184 p.

Singh, Y.D., J. Joshua, S.F.W. Sunderraj, V.V. Kumar, O. Deiva, P.N. Joshi, N.M. Joshi and H. Soni (2003) Rare and Endangered Plants and Animals of Gujarat. Under Save Biodiversity Campaign, Sponsored By Gujarat Ecology Commission (GEC), Vadodara. 79 pp.

Smith, C.F. (1978) Distributional Ecology of Barred and Great Horned Owls in Relations to Human Distribution. M.S. Thesis. University of Connecticut.

Smith, D.G, Ellis, D.H., Millsap, B.A. (1988) Owls. In: Proceedings of the Southeast Raptor Management Symposium and Workshop. National Wildlife Federation, Washington, D.C. National Wildlife Scientific and Technical Series, no. 14. Pp. 89-117.

Smith, D.L. (1997) Folklore of the Winnebago Tribe. Norman: University of Oklahoma Press, p. 160.

Smythies, B.E. (1953) The Birds of Burma. 2nd edition. [Publisher unknown], Edinburgh-London.

Sodoma, B. (1999) Builders Awaiting Future of Pygmy Owl Habitat. Inside Tucson Business. 9 (18): 78.

Soni, H. (2001) Birds seen in and around Gomti Pond, Dakor. Vihang. P. 22.

Soni, H. (2001) Pelicans in Anand Territory. Vihang. P. 15.

Soni, H. (2001) White-throated Munia. Vihang. 21-22.

Soni, H. (2002) A Birding Note. Vihang. P. 17.

Soni, H. (2002) Birds seen in and around Kapadvanj. Vihang. P. 12.

Soni, H. (2002) Birds seen in and around Madhasna, Kheralu Taluka, Mehsana District. Vihang. P. 12.

Soni, H. (2002) Some Avifaunal Observations in North Gujarat (Banaskantha). Vihang. 12-13.

Soni, H. (2003) Some Miscellaneous Birding Observations. Vihang. P. 21.

Soni, H. (2004) A Survey of White-backed Vulture Gyps bengalensis in Gujarat. In: Inputs from Birdwatchers for Workshop on "Current Status of Vultures in Gujarat". (Eds. Jani, J.J., H.R. Kher, D.J. Patel, A. Tere, D.N. Rank). Bird Conservation Society Publication, Ahmedabad, Gujarat. 108 pp.

Soni, H. (2004) Mysterious Death. Down To Earth. 3-4. (ISSN: 0012-5792)

Soni, H. (2006) Grey-headed Flycatcher in Outskirts of Jamnagar City, Gujarat. Flamingo. (Newsletter of Bird Conservation Society of Gujarat). 4 (1 & 2): 13.

Soni, H. (2007) A Survey of White-rumped Vultures Gyps bengalensis in Gujarat State, India. Vulture News. 57: 32-43. (ISSN: 1606-7479) (South Africa)

Soni, H. (2007) Mass Mortality of Sea Gulls at Lakhota Lake, Jamnagar, Gujarat. Flamingo. (Newsletter of Bird Conservation Society of Gujarat). 5 (1 & 2): 5-6.

Soni, H. (2007) Prevalence of Some Mythological Beliefs among Rural Communities of Gujarat: A Case Study of Crow (Corvus sp.). Newsletter for Birdwatchers. 47 (3): 48. (ISSN: 0028-9426)

Soni, H. (2008) Regional Names of Birds of Nal Sarovar Bird Sanctuary, Gujarat – A Case Study. Vihang. 2: 28-31.

Soni, H. (2009) Mass Mortality of Pariah Kite (Milvus migrans) in Ahmedabad Sewage Treatment Plant, Gujarat. Newsletter for Birdwatchers. 49 (2): 26-29. (ISSN: 0028-9426)

Soni, H. (2009) Regional Names of Birds of Purna Wildlife Sanctuary (Dangs Forest), Gujarat – A Case Study. Vihang. 1: 38-39.

Soni, H. (2010) Calling Pattern in Coppersmith Barbet (Megalaima haemacephala). Newsletter for Birdwatchers. 50 (1): 2-6. (ISSN: 0028-9426)

Soni, H. (2013) Wilderness and Forests: Our Last Bastions of Biodiversity. Guide.Net Newsletter. 2 (5): 4-7.

Soni, H. (2014) About the Directory of Birdwatchers of Gujarat State. Newsletter for Birdwatchers. 54 (3): 35. (ISSN: 0028-9426)

Soni, H. (2014) About the Directory of Birdwatchers of Gujarat State. Newsletter for Birdwatchers. 54 (3): 35. (ISSN: 0028-9426)

Soni, H. and J. Joshua (2009) A Clarification regarding the Sighting of Yellow-bellied Prinia (Prinia flaviventris Delessert, 1840) – First Record in Gujarat. Newsletter for Birdwatchers. 49 (1): 1-2. (ISSN: 0028-9426)

Soni, H. and J. Joshua (2012) Crested Bunting Melophus lathami (Gray) – A First Report for Ambakantha Forest (Banaskantha District), North Gujarat. Newsletter for Birdwatchers. 52 (5): 76. (ISSN: 0028-9426)

Soni, H. and J. Joshua (2012) First Report of Jungle or Grey Nightjar (Caprimulgus indicus Latham) in Khadir Island, Rapar (Kachchh), Gujarat. Newsletter for Birdwatchers. 52 (6): 87. (ISSN: 0028-9426)

Soni, H. and J. Joshua (2012) Red-necked Phalarope (Phalaropus lobatus, Linnaeus, 1758) in Breeding Plumage in Freshwater Ecosystem of Kachchh District, Gujarat. Newsletter for Birdwatchers. 52 (6): 93. (ISSN: 0028-9426)

Soni, H. and J. Joshua (2012) Use of Concrete Electricity Pole by Yellow-fronted Pied Woodpecker (Picoides mahrattensis Latham). Newsletter for Birdwatchers. 52 (4): 60-61. (ISSN: 0028-9426)

Soni, H. and J. Joshua (2013) Breeding of Rock Bunting (Emberiza cia Linnaeus, 1766) in Kachchh, Gujarat.

Newsletter for Birdwatchers. 53 (3): 46-47. (ISSN: 0028-9426)

Soni, H. and J. Joshua (2013) Sighting of Rufous-tailed Scrub Robin (Cercotrichas galactotes Temminck, 1820) in Pachchham Island, Kachchh District, Gujarat. Newsletter for Birdwatchers. 53 (4): 62. (ISSN: 0028-9426)

Soni, H. and J. Joshua (2013) Sighting of Spotted Creeper (Salpornis spilonotus Franklin, 1831) in Balaram-Ambaji Wildlife Sanctuary, North Gujarat. Newsletter for Birdwatchers. 53 (4): 63. (ISSN: 0028-9426)

Soni, H. and J. Joshua (2013) Verditer Flycatcher (Muscicapa thalassina Swainson, 1838) (Passeriformes: Muscicapidae) in Kachchh (Gujarat). Newsletter for Birdwatchers. 53 (3): 42. (ISSN: 0028-9426)

Soni, H. and J. Joshua (2014) Spotted Flycatcher (Muscicapa striata Pallas, 1764) in Jessore Sloth Bear Sanctuary, North Gujarat. Newsletter for Birdwatchers. 54 (5): 57. (ISSN: 0028-9426)

Soni, H., J. Pankaj and J. Justus (2005) Nesting Behaviour and Unusual Feeding Pattern in Common Wood Shrike (Tephrodornis pondicerianus). Journal of Bombay Natural History Society. 102 (1): 120. (ISSN: 0006-6982)

Soni, H., P. Joshi and J. Joshua (2003). Breeding of White-backed Vulture (Gyps bengalensis) in Kachchh. Flamingo. (Newsletter of Bird Conservation Society of Gujarat). 1 (5 & 6): 8.

Soni, H.B. (2004) A Red Alert: Bird's Trade. Flamingo. (Newsletter of Bird Conservation Society of Gujarat). 2 (1 & 2): 9.

Soni, H.B. (2011) Biodiversity: A Burgeoning Biospectrum of Life. Quest (ARIBAS Newsletter). 1 (2): 6-7.

Soni, H.B. (2011) Biological Diversity of Gujarat: A Holistic Scenario. Quest (ARIBAS Newsletter). 1 (1): 7-9.

Soni, H.B. (2011) Policy Framework: A Legislative Tool for Biodiversity Protection. Quest (ARIBAS Newsletter). 1 (3): 8-9.

Soni, H.B. (2014) Chronological Perspectives and Bibliographical Milestones in Ornithology of Gujarat State, India: A Five Years Journey (2010-2014) – Part I. Newsletter for Birdwatchers. 54 (4): 43-46. (ISSN: 0028-9426)

Soni, H.B. (2014) Chronological Perspectives and Bibliographical Milestones in Ornithology of Gujarat State, India: A Five Years Journey (2010-2014) – Part II. Newsletter for Birdwatchers. 54 (5): 52-56. (ISSN: 0028-9426)

Soni, H.B. (2015) Stoliczka's Bushchat in Kachchh. Hornbill. 1 (January-March): 22.

Soni, H.B. (2016) A Review on Birds of Arid Regions of North-Western India. Guide.Net Newsletter. 5 (2): 4-7.

Soni, H.B. (2017) Birds in Ancient Times: Myths and Motifs. Newsletter for Birdwatchers. 57 (6): 67-70. (ISSN: 0028-9426)

Soni, H.B. (2017) Wetlands – An Eternal Abode for Avifauna. India Wilds Newsletter. 9 (4): 17-28. (ISSN: 2394-6946)

Soni, H.B. (2018) An Integral Association between Trees and Birds. Newsletter for Birdwatchers. 58 (2): 19-23. (ISSN: 0028-9426)

Soni, H.B. (2018) Bird Migration: A Peep into Past and Present. Newsletter for Birdwatchers. 58 (6): 64-70. (ISSN: 0028-9426)

Soni, H.B. (2018) Magnetic Orientation in Birds: A Miraculous Mechanism (Part 1). Newsletter for Birdwatchers. 58 (3): 34-35. (ISSN: 0028-9426)

Soni, H.B. (2018) Magnetic Orientation in Birds: A Miraculous Mechanism (Part 2). Newsletter for Birdwatchers. 58 (4): 39-45. (ISSN: 0028-9426)

Soni, H.B. (2020) Biodiversity and Birds: A Field Journey (Amazon Kindle Edition). Kindle Direct Publishing (KDP), USA. 147 pp. (ASIN: B08HMC789R) https://www.amazon.in/dp/B08HMC789R

Soni, H.B. (2020) Biodiversity and Birds: A Field Journey (Amazon Paperback Edition). Kindle Direct Publishing (KDP), USA. 199 pp. (ISBN-13: 979-8691628733; ASIN: B08KBV5CX6) https://www.amazon.com/dp/B08KBV5CX6

Soni, H.B. (2020) Biodiversity of Wetlands and Forests: A Nature Trail (Google Play Books Edition). Google Book Publisher (GBP), USA. 165 pp. (GGKEY: ZEK5C67C9E6)

https://play.google.com/store/books/details?id=gAf-DwAAQBAJ

Soni, H.B. (2020) Biodiversity, Birds and Wetlands (Amazon Kindle Edition). Kindle Direct Publishing (KDP), USA. 109 pp. (ASIN: B08HBMGSGD) https://www.amazon.in/dp/B08HBMGSGD

Soni, H.B. (2020) Biodiversity, Birds and Wetlands (Amazon Paperback Edition). Kindle Direct Publishing (KDP), USA. 151 pp. (ISBN-13: 979-8685500144; ASIN: B08KBGMG26)
https://www.amazon.com/dp/B08KBGMG26

Soni, H.B. and U. Kathad (2014) Photographic Record of Opportunistic Feeding by Great White Pelican (Pelecanus onocrotalus) on Little Cormorant (Phalacrocorax niger). Newsletter for Birdwatchers. 54 (3): 32-35. (ISSN: 0028-9426)

Soni, H.B., P.N. Joshi, J. Joshua and S.F.W. Sunderraj (2016) In Search of Stoliczka's Bushchat (Saxicola macrorhyncha Stoliczka, 1872) in Kachchh Arid Ecosystem, Gujarat. India Wilds Newsletter. 8 (3): 7-10. (ISSN: 2394-6946)

Sparks, J., Soper, T. (1979) Owls: Their Natural and Unnatural History. p. 163. ISBN 0715349953

Sparks, J., Soper, T. (1970) Owls: Their Natural and Unnatural History. Taplinger Publishing Company, New York.

Stikini, an owl monster of Seminole folklore. Native-languages.org.

Stokes, D.W., Stokes, L.Q. (1996) Stokes Field Guide to Birds: Eastern Region. Little, Brown and Company, Boston.

Take A Peek At Boo, The Eagle Owl – The Quillcards Blog. Quillcards.com

The Hungry Owl Project. Hungryowl.org.

The Popol Vuh (2008) Meta-religion.com.

The Significance and Meaning of Owls in Japanese Culture. Owlcation.com

Thiselton-Dyer, T.F. (1883) Folklore of Shakespeare. Sacred-texts.com

Tischendorf, J., Johnson, C.L. (1997) Long-Eared Owl Snagged on Barbed-Wire Fence. Blue Jay. 55: 200.

Voous, K.H. (1988) Owls of the Northern Hemisphere. The MIT Press, Cambridge, Mass. 320 pp.

Wallace, D.R. (1983) The Klamath Knot. Sierra Club Books, San Francisco, California. 149 pp.

Walls, G.L. (1942) The vertebrate eye and its adaptive radiation. Cranbook Institute of Science.

Webster, D.B.F., Richard, R. (2012) Hearing in Birds. The Evolutionary Biology of Hearing. Springer Science & Business Media. p. 547. ISBN 978-1-4612-2784-7.

Weidensaul, S. (1996) Raptors: The Birds of Prey: An Almanac of Hawks, Eagles and Falcons of the World. Lyons and Burford, New York, NY.

Weinstein, K. (1989) The Owl in Art, Myth and Legend. Crescent Books, N.Y.

Wildlife Trade News – Huge haul of dead owls and live lizards in Peninsular Malaysia. TRAFFIC. 12 November 2008.

Willott, J.F. (2001) Handbook of Mouse Auditory Research. CRC Press. ISBN 1420038737.

Work, T.M., Hale, J. (1996) Causes of Owl Mortality in Hawaii, 1992-1994. Journal of Wildlife Diseases. 32 (2): 266-273.

www.ingramcontent.com/pod-product-compliance
Lightning Source LLC
LaVergne TN
LVHW091232180726

843490LV00006B/2052